CAMBRIDGE LIBRARY COLLECTION

Books of enduring scholarly value

Technology

The focus of this series is engineering, broadly construed. It covers technological innovation from a range of periods and cultures, but centres on the technological achievements of the industrial era in the West, particularly in the nineteenth century, as understood by their contemporaries. Infrastructure is one major focus, covering the building of railways and canals, bridges and tunnels, land drainage, the laying of submarine cables, and the construction of docks and lighthouses. Other key topics include developments in industrial and manufacturing fields such as mining technology, the production of iron and steel, the use of steam power, and chemical processes such as photography and textile dyes.

A Handbook of Wireless Telegraphy

James Erskine-Murray (1868–1927) was a Scots expert in wireless technology who studied under Lord Kelvin for six years at Glasgow University before arriving at Trinity College, Cambridge as a research student. He eventually became a telegraphy consultant and published this work in 1907. Its aim was to inform engineers, students, and radio operators about many aspects of a rapidly changing technology. The book covers recent developments of the time, and a whole chapter is dedicated to the issue of transmission. Erskine-Murray also provided a chapter of tables containing data which he calculated himself and which had not appeared in print before. The work stands as a classic in the field of early engineering texts, and offers contemporary students and radio enthusiasts a useful guide to early wireless technology.

Cambridge University Press has long been a pioneer in the reissuing of out-of-print titles from its own backlist, producing digital reprints of books that are still sought after by scholars and students but could not be reprinted economically using traditional technology. The Cambridge Library Collection extends this activity to a wider range of books which are still of importance to researchers and professionals, either for the source material they contain, or as landmarks in the history of their academic discipline.

Drawing from the world-renowned collections in the Cambridge University Library, and guided by the advice of experts in each subject area, Cambridge University Press is using state-of-the-art scanning machines in its own Printing House to capture the content of each book selected for inclusion. The files are processed to give a consistently clear, crisp image, and the books finished to the high quality standard for which the Press is recognised around the world. The latest print-on-demand technology ensures that the books will remain available indefinitely, and that orders for single or multiple copies can quickly be supplied.

The Cambridge Library Collection will bring back to life books of enduring scholarly value (including out-of-copyright works originally issued by other publishers) across a wide range of disciplines in the humanities and social sciences and in science and technology.

A Handbook of Wireless Telegraphy

Its Theory and Practice, for the Use of Electrical Engineers, Students, and Operators

James Erskine-Murray

CAMBRIDGE UNIVERSITY PRESS

Cambridge, New York, Melbourne, Madrid, Cape Town, Singapore,
São Paolo, Delhi, Dubai, Tokyo, Mexico City

Published in the United States of America by Cambridge University Press, New York

www.cambridge.org
Information on this title: www.cambridge.org/9781108026888

This edition first published 1907
This digitally printed version 2011

ISBN 978-1-108-02688-8 Paperback

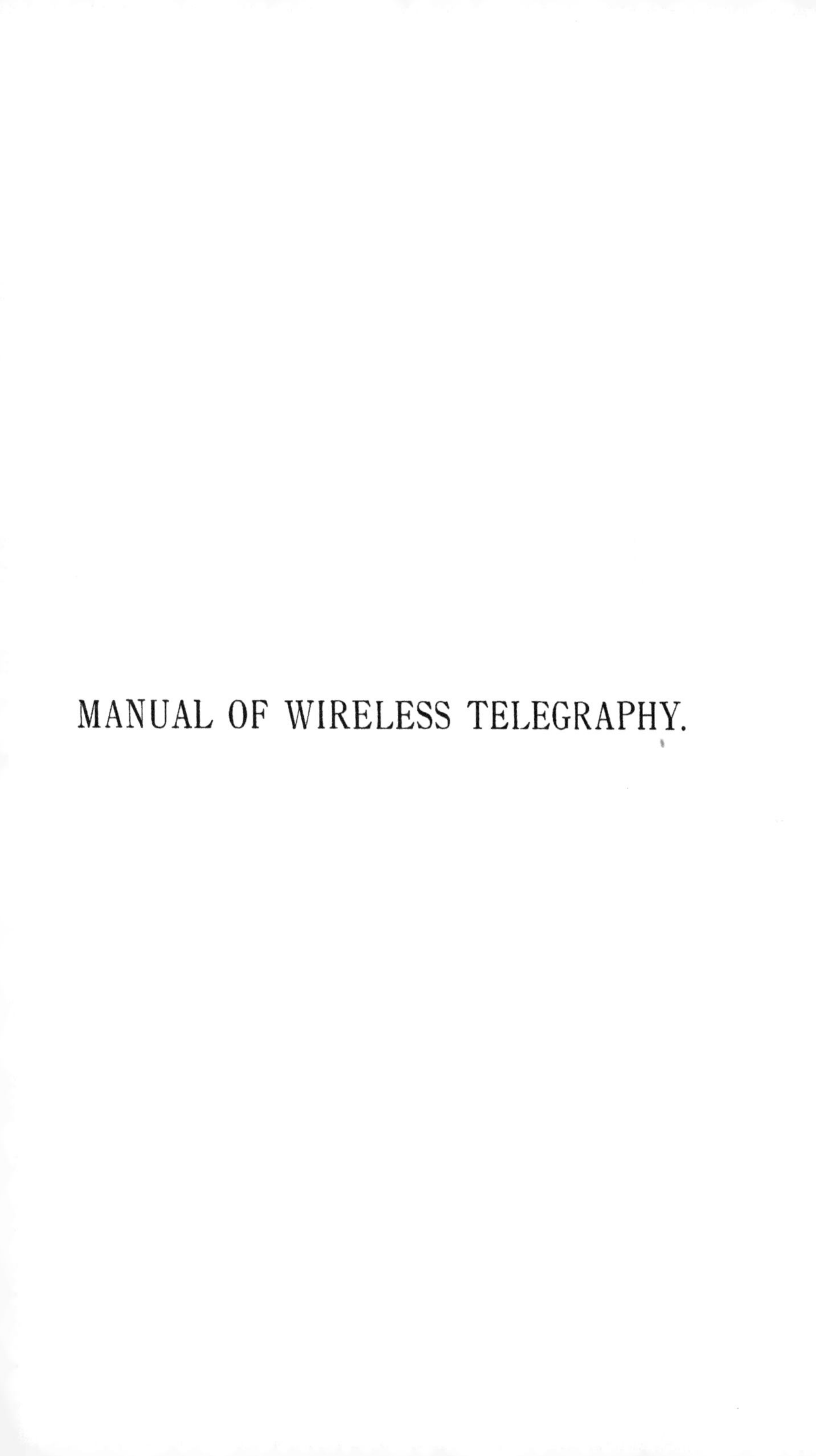

MANUAL OF WIRELESS TELEGRAPHY.

A HANDBOOK OF WIRELESS TELEGRAPHY

Its Theory and Practice

FOR THE USE OF ELECTRICAL ENGINEERS, STUDENTS, AND OPERATORS

BY

JAMES ERSKINE-MURRAY, D.Sc.

FELLOW OF THE ROYAL SOCIETY OF EDINBURGH
MEMBER OF THE INSTITUTION OF ELECTRICAL ENGINEERS

LONDON
CROSBY LOCKWOOD AND SON
7 STATIONERS' HALL COURT, LUDGATE HILL
1907

This Book is dedicated

TO

THE RIGHT HONOURABLE

LORD KELVIN, O.M., G.C.V.O.

WHOSE INSPIRING TEACHING AND
SYMPATHETIC INTEREST HAVE
BEEN THE SOURCE OF
WHATEVER SCIENTIFIC
MERIT IT MAY
POSSESS.

PREFACE.

THIS "Handbook of Wireless Telegraphy" (it need hardly be said) is not encyclopædic; nor, on the other hand, does it pretend to be a simple exposition of its subject for the benefit of the uninitiated. Rather it is intended for the use of those who, for reasons of business or pleasure, have already made themselves acquainted with at least some of the truths of the theory and practice of wireless telegraphy, and to whom, therefore, the ordinary technical terms will convey something more concrete than a mere definition in words.

Few, if any, quotations are made in the volume which do not bear directly on wireless telegraphy, and those which are inserted are given fully. The reader should, therefore, have sufficient material on which to form his own estimate of the practical bearing of some of the more important researches of earlier workers in this subject.

The thanks of the Author are due to the Companies and Firms who have kindly supplied him with details of their operations and with blocks illustrative of their systems, thus enabling him to give, also, authoritative descriptions of the latest developments. Statements of achievements have been backed, as far as possible, by reports of official experts, but failing this, results have, as a rule, been given on the authority of the inventors or of the assignees of their patents.

In Chapter XVII. will be found the Author's views on the subject of the mechanism of Transmission. These are based on the latest researches; and although as yet the theory is far from complete, it will be seen that some advance has been made through the consideration of actual outdoor circumstances, in addition to results of experimental measurements.

The Tables in Chapter XXI. form a unique feature of the volume, and should render it useful to those engaged in the practice of wireless telegraphy. Most of them have been calculated by the Author and appear in print for the first time. They deal with such data as the inductance of helices, capacity of condensers of various forms and sizes, resistance of various conductors to high frequency currents, and similar numerical quantities required in the design of radio-telegraphic apparatus. As the calculation of these Tables has involved a large amount of arithmetical work, it is possible that (in spite of precautions) some errors may have crept in, and the Author would be grateful if any reader who may discover any error will kindly communicate with him through the Publishers.

Chapter XX. is an almost literal translation of a paper "On the Calculation of a Syntonic Wireless Telegraph Station," by Signor Alfredo Montel, which was published at Rome in June 1906. It gives concisely, with numerical examples, the calculations necessary for the design of a modern wireless telegraph station. The Author's grateful thanks are due to Signor Montel, and to the proprietors of *L'Elletricista* for their kind permission to reproduce here in English dress a translation of this most useful and important paper.

From what has been said, it will be seen that the Author has tried to confine himself strictly to modern open circuit wireless telegraphy, and has omitted methods of communication which do not come under that description. In regard to mathematical theory, he has avoided, as far as possible, the discussion of any problem which remains to be dealt with more satisfactorily by experiment. In regard to the relative merits of close and loose coupling, and many other questions arising in practical work, it must be remembered that, with the instruments now at the command of wireless telegraph engineers, it is possible in the course of a brief experiment to obtain comparative results which are as satisfactory as a mathematical demonstration could be.

It should be mentioned that the Author has taken it upon himself to introduce one or two new terms—such as "jig," "equifrequent," and "revibration." The first is short for "an alternating electric current of the high frequency usually adopted in wireless telegraphy;" while the others are almost self-explanatory, "revibration" describing in general the action of which resonance is a particular example.

In addition to acknowledgments already made, the Author must express his grateful thanks to the authors and publishers who have assisted him by supplying information, or by granting permission for the use of extracts or illustrations from published matter. In particular he has to thank the proprietors of the *Electrician*, without whose generous aid it would have been impossible for him to have included authorised or full descriptions of many recent discoveries; as well as the Royal Societies of London and Edinburgh, the Royal Physical Society, the Institution of Electrical Engineers, and the proprietors of the *Philosophical Magazine*, the *Electrical World and Engineer* (*New York*), and the *Electrical Review*. He is likewise indebted to Messrs Longmans and Professor Fleming for blocks illustrative of the Marconi system; to Messrs Macmillan and his friend Professor Gibson for an extract from the latter's treatise on "Graphs"; to Mr Commerford Martin for extracts from his "Inventions of Nikola Tesla;" and to Dr Barth, of Leipsic, for blocks taken from Hertz's "Electric Waves."

Finally, he must express his thanks to his friend Professor Magnus Maclean for kindly checking the Tables of Chapter XXI., and to Professor Andrew Jamieson for many useful hints during the progress of the work.

LONDON, *February* 1907.

CONTENTS.

CHAPTER VI.

CHAPTER VII.

CHAPTER VIII.

CHAPTER IX.

CHAPTER X.

CHAPTER XI.

CHAPTER XII.

CHAPTER XIII.

CHAPTER XIV.

CHAPTER XV.

CHAPTER XVI.

CHAPTER XVII.

CHAPTER XVIII.

CHAPTER XIX.

CHAPTER XX.

CHAPTER XXI.

SELECTED LIST OF ILLUSTRATIONS.

A HANDBOOK OF

WIRELESS TELEGRAPHY.

CHAPTER I.

ADAPTATIONS OF THE ELECTRIC CURRENT TO TELEGRAPHY.

PROGRESS in the branch of electric telegraphy usually called "wireless" has been so very rapid that many of the technical terms used have not yet had definite meanings attached to them. To begin with, the term "wireless" itself is used vaguely to cover many systems of totally different kinds which have in common only the fact that no insulated conductor joins the sending and receiving stations. The use of the word is no doubt correct, but it leaves undefined the type of apparatus and method of transmission.

Some writers attempt to distinguish the more modern methods as "electric wave telegraphy," "space telegraphy," "Hertzian telegraphy," "spark telegraphy," and so on; but few if any of these designations really go to the root of the matter, and many of them might equally well be applied to systems totally unlike those which they are meant to describe.

Let us consider the course of the electric current in various telegraphic systems.

In ordinary wire telegraphy the current goes by the whole Earth and returns by the wire. Telegraphists usually put it the other way, but it is a matter of indifference. The circuit is a closed conducting loop, and, barring unintended leakage, the whole current sent out passes through the receiver.

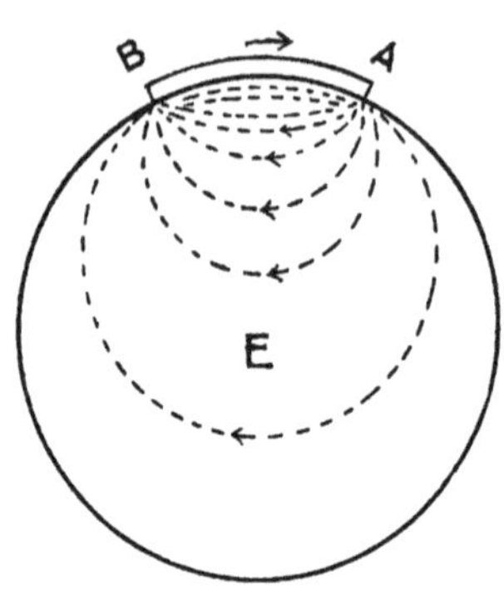

FIG. 1.—WIRE TELEGRAPH.
E, the Earth; *A B*, Insulated Wire connected to Earth through Sending and Receiving Instruments at *A* and *B*. Arrowheads show direction of current.

In the Morse, Lindsay, and Willoughby-Smith systems a current is made to flow between two plates buried in the earth or sea, and the receiver current is picked up by other earth-plates placed on the lines of flow of the current. The conducting circuit is still a completely closed one. In fact, it exactly resembles a shunted galvanometer. The earth is the shunt, hence only

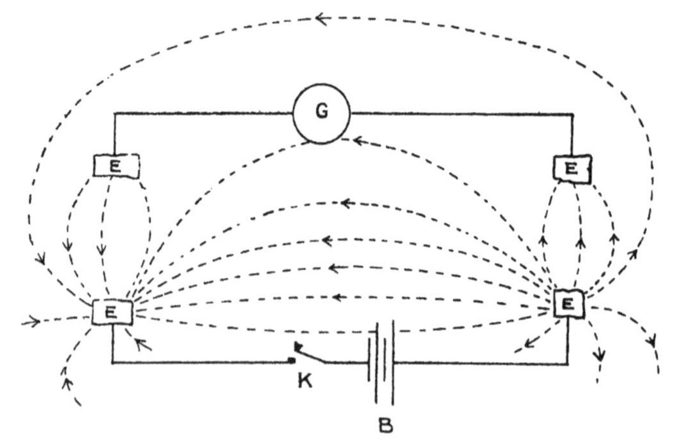

FIG. 2.—GENERAL DIAGRAM OF CLOSED CIRCUIT WIRELESS TELEGRAPH (Morse, Lindsay, Willoughby-Smith).

E, Earth-Plates; *K*, Transmitting Key; *B*, Source of Current; *G*, Receiving Galvanometer. Arrowheads show direction of current in earth or sea.

a small proportion of the transmitted current flows through the receiver.

One of Sir Oliver Lodge's systems, not that in most general use, depends on varying magnetic induction through two closed conducting circuits, one at the sending and the other at the receiving station. No current of electricity passes from one to the other, but only the magnetic force.

Between these extremes stand the systems which have been most successful. In all of them a conduction current, not continuous, but alternating, passes through the upper layers of the earth's surface from the sending to the receiving station. As, however, there is no return wire the current spreads, as a rule, equally in all directions round the transmitter.

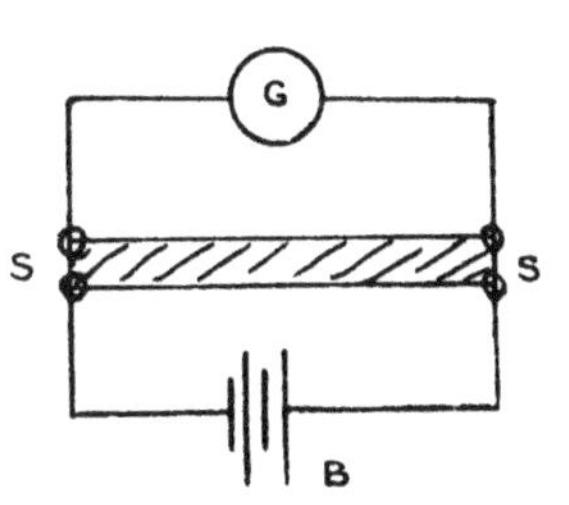

FIG. 3.—SHUNTED GALVANOMETER. (Compare with Fig. 2.)

B, Source of Current; *S S*, Thick Conductor used as Shunt; *G*, Galvanometer.

Reduced to its simplest terms the modern wireless telegraph is a large conducting sphere (the earth) with two conducting excrescences on it or near its surface (the aerial conductors). In one of these a sudden oscillatory movement of electricity is started, which spreads over the surface, causing to-and-fro currents in the other wire as it passes.

In several systems no conducting connection is made between the aerials and the earth. In such cases the current becomes in places a dielectric displacement. It is, however, a true current in the general sense, and the configuration of its lines of flow is similar to that in the entirely conducting systems, except that there may be in addition a set of closed lines of force or direct radiation propagated outwards above the earth, as in the Lodge-Muirhead system.

So far we have considered mainly the conduction current. There is always associated with it a steady or varying displacement in the dielectric or insulator surrounding it—in our case in the atmosphere. In wireless telegraphy this displacement is rapidly alternating and in itself constitutes a dielectric current, or perhaps we should rather say, currents, since the lines of force do not, except in the case of stations within short distances of one another,

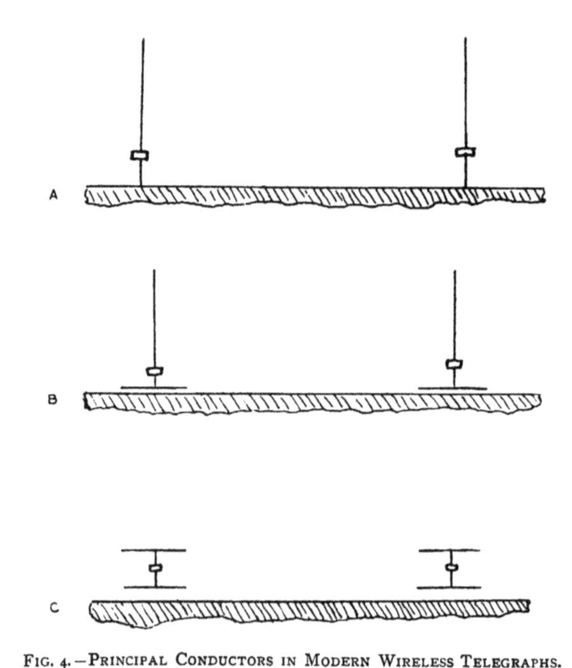

FIG. 4.—PRINCIPAL CONDUCTORS IN MODERN WIRELESS TELEGRAPHS.
A, Aerial Wires connected conductively to Earth through Transmitting and Receiving Instruments; *B*, Aerial Wires connected inductively to Earth by means of Wire Netting spread on ground; *C*, Lodge-Muirhead system.

bridge directly from one aerial to the other. Of course, if the first process of charging the sending aerial is comparatively slow, taking, for instance, as much as a thousandth of a second, the lines of force will have time to spread a great distance, and the electrification on the aerials and earth's surface will have attained a practically static condition before any further change occurs. The lines of force move outward at about 186,000 miles per second, so that in one-thousandth of a second the far ends of the outermost ones will, where they touch the earth, have reached a distance of 186 miles, the near ends still remaining on the sending aerial. Thus we see why it is possible to obtain a signal at many miles' distance when the primary circuit is broken, even though no spark occurs in the transmitter and no oscillations take place.

In systems in which the aerial does not contain a spark-gap, but is coupled by induction to a separate oscillating circuit, this initial slow charge is probably quite negligible; it is certainly evanescent. Most systems now in use con-

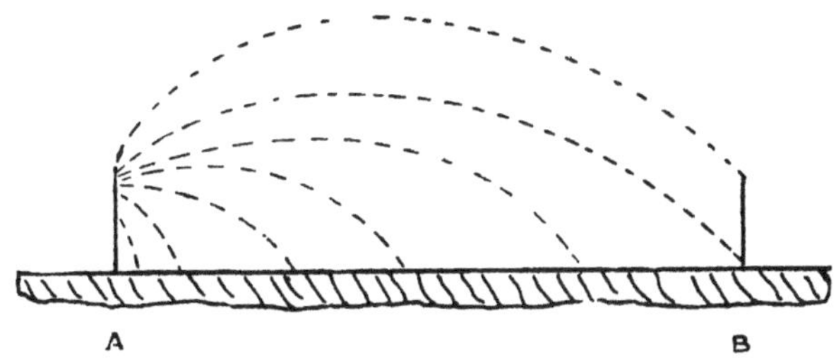

FIG. 5.—DISTRIBUTION OF USEFUL LINES OF ELECTRIC FORCE BEFORE SPARK OCCURS WHEN DISTANCE SMALL.
A, Transmitting Aerial; *B*, Receiving Aerial.

form to these conditions, and therefore do not commence every wave train with an odd wave of great length, but start with one of the same length as those which follow it.

Modern wireless telegraphy is, in general, open circuit telegraphy, *i.e.*, telegraphy from one part of a conductor to another without the use of a return wire.

True Hertzian telegraphy, with small radiators placed more than a wave length above the ground, uses no continuous conductor between the stations. *Short-wave* Hertzian telegraphy has never been successful at distances beyond a mile or two, and apparently cannot compete with systems which utilise the earth as conductor to guide the current round corners, or with long-wave systems in which diffraction performs the same function. As a matter of fact, all the systems now in use have either direct conducting connections with the earth or have the lower end of the vertical aerial so near the ground that the condenser formed by the two is nearly as capable of carrying the oscillating current as a conductor would be.

I shall leave questions of priority of invention to historians and lawyers, and content myself with describing only the experiments which appear to have contributed most directly to the development of the new means of communication. It should therefore be understood that where dates are mentioned they apply only to the matter in hand and do not imply the commencement of the work of any particular experimenter or even his first publication on the subject.

Closed Circuit Wireless Telegraphy.—The earlier systems * of electric telegraphy without connecting wires such as those of Morse and Lindsay depended on the detection of minute differences of potential between two earth-plates sunk at some distance apart on a line flow of the current between the two earth-plates of the transmitter. These lines of flow resemble, in form, the lines of force between the poles of a magnet, and as in the case of the magnet the intensity of the action decreases very rapidly with distance from the source. This method was therefore of very limited application, for it entailed the use of at least twice as much wire as would have served to connect the stations directly, had that been possible. The long base lines required between the earth-plates at both the transmitting and receiving stations precluded entirely its use for communication between ship and shore. The Armstrong and Orling system is the only modern one which depends on this principle, and, though pairs of earth-plates are used, the transmitting and receiving instruments

* See "History of Wireless Telegraphy," by J. J. Fahie.

have been so much developed in power and sensibility that the ratio of the distance between transmitter and receiver to the distance between either pair of *connected* earth-plates has been enormously increased.

Sir W. Preece, also, employs earth-plates at the ends of elevated horizontal wires, and though the actual process of transmission is somewhat different its limitations are similar. This system has, however, been of use under certain circumstances, and there are a few installations working in the United Kingdom.

Open Circuit Wireless Telegraphy.—Leaving then these systems in which the lines of electric force are moving at right angles to the direction of transmission, and in which the intensity of their action therefore falls off approximately as the cube of the distance, let us consider the more modern systems in which the lines of force move in the direction of transmission, and have an intensity which decreases in simple proportion to the distance. This difference constitutes in itself so enormous an advantage in favour of the single earth connection and vertical wire as opposed to the long horizontal wire earthed at both ends, that the progress of systems depending on the latter has been entirely eclipsed by the development of the former. In this book I shall therefore deal almost exclusively with the types of apparatus based on the work of Hertz, Lodge, Jackson, Trouton, Marconi, and their many followers, and not with wireless telephony, or with any closed circuit system.

We must now consider the experimental discoveries previous to the invention of modern wireless telegraphy, but on which it is based. These may be conveniently divided into (1) those which rendered the production and transmission of electric oscillations possible; and (2) those which made the detection of these oscillations easy. In the former category the researches of Henry, Von Bezold, Hertz, Lodge, and Tesla stand pre-eminent, and in the latter the principal workers were Hughes, Calzecchi-Onesti, Branly, Lodge, and Rutherford.

Henry.—Professor Joseph Henry, of Princeton University, was the first to show experimentally that the discharge of a Leyden jar may be oscillatory in character, a discovery which bears directly on the invention of wireless telegraphy, not only on account of the production of a succession of waves, but also because the changes of potential are so much more rapid that in a steady discharge. In 1838 Henry even succeeded in demonstrating that the oscillations in a circuit containing a Leyden jar and spark-gap would induce currents in an independent circuit many feet distant. Electric telegraphy was, however, in its infancy, and no attempt was made to adapt the apparatus to the needs of practical communication. Henry's observations on the oscillatory character of the discharge were confirmed experimentally by Helmholtz and Feddersen, and their theory was given by Lord Kelvin (see Chapter XVI.).

Professor D. E. Hughes.—In 1879 Hughes showed that it was possible to transmit signals without the use of a connecting wire, from an induction coil to a microphone several hundred yards away. He rightly believed that the oscillations were propagated through the insulating medium between the instruments, but discouraged by the scepticism of several well-known persons who formed a committee of inquiry, he did not continue his work beyond the purely experimental stage. The facts, though known to the committee, were not published until over twenty years later, when Marconi had already brought out his well-known system.

Sir Oliver Lodge.—While studying the nature of the spark discharge from condensers, Sir Oliver Lodge * dis-

* *Nature*, vol. 41, p. 368, Feb. 1890.

covered that it was possible, by equalising the electrical dimensions of two circuits, to obtain an electrical effect corresponding to the resonance of two mechanical vibrators of equal frequency. The phenomenon is of great importance in modern wireless telegraphy, and though in its original form the apparatus used did not suggest a practical means of communication, a similar arrangement of mutual equifrequent circuits forms an important part of almost every wireless telegraph instrument. It is therefore worth while to consider carefully the actual experiments as originally carried out. I shall therefore quote Sir Oliver Lodge's description :—

"I shall show this [electrical resonance] in a form which requires great precision of tuning or syntony, both emitter and receiver being persistently vibrating things giving some thirty or forty swings before damping has a serious effect. I take two Leyden jars with circuits about a yard in diameter, and situated about two yards apart (Fig. 6). I charge and discharge one jar, and observe that the surgings set up in the other can cause it to overflow if it is syntonised with the first (see *Nature*, vol. XLI. p. 368 ; J. J. Thomson, 'Recent Researches,' p. 395). A closed circuit such as this is a feeble radiator and a feeble absorber, so it is not adapted for action at a distance. In fact I doubt whether it will visibly act at a range beyond the $\frac{1}{4}\lambda$ at which true radiation of broken-off energy occurs. If the coatings of the jar are separated to a greater distance, so that the dielectric is more exposed, it radiates better; because in true radiation the electrostatic and the magnetic energy are equal, whereas in a ring circuit the magnetic energy greatly predominates. By separating the coats of the jar as far as possible we get a typical Hertz vibrator (Fig. 6A), whose dielectric extends out into the room, and thus radiates very powerfully."

As means of communication, systems based on the interaction of two closed resonating circuits cannot compete favourably with open circuit systems, for the energy transmitted shows an exceedingly rapid decrease with distance. The fact that the possibility of producing electrical resonance had been experimentally demonstrated was, however, of fundamental importance, and opened up the way for further experiment. Many peculiarities of

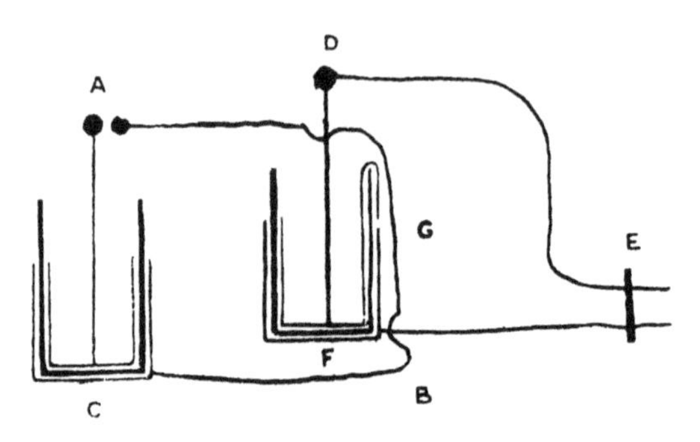

FIG. 6.—LODGE'S SYNTONIC LEYDEN JAR CIRCUITS.

A, *B*, *C*, Transmitting Circuit; *D*, *E*, *F*, Receiving Circuit; *A*, Spark-Gap; *E*, Slider for Tuning by altering Receiving Circuit; *G*, Small Overflow Spark-Gap for detecting Received Current.

alternating currents of very high frequency were first pointed out by Lodge. Following an observation of Faraday's in a simple yet remarkable experiment, he showed that the self-induction of a single loop of thick wire a few feet in diameter, was so great a hindrance to an oscillating current

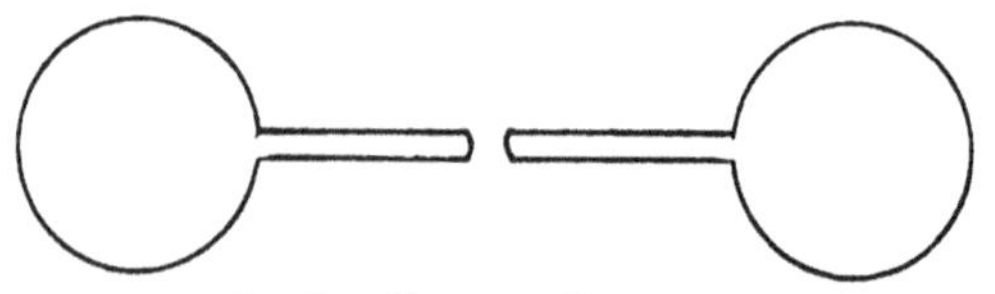

FIG. 6A.—HERTZIAN OSCILLATOR.

of high frequency that a spark of considerable length, indicating a difference of potential of thousands of volts, passed across from one point on the loop to another, though the conductor forming the loop had a resistance of only a small fraction of an ohm.

The enormous importance of the inductance of any circuit along which it is proposed to send high frequency currents can hardly be overestimated by those accustomed to work with currents of low frequency.

A few examples of the calculation of the impedance of a circuit for currents of different frequencies will be found in the tables at the end of this book.

Capacity plays an equally important part in wireless telegraphy, and is never a negligible factor. Indeed most problems in the subject resolve themselves into questions concerning the capacity, inductance, and therefore the geometrical form of the conductors concerned. The material of the conductor is generally of secondary importance, but its size and shape and the dielectric properties of the insulating material surrounding it determine the distribution of the lines of electric force, and, if these are in motion, the system of magnetic lines which constitutes the self-induction or inertia of the circuit. The actual form of the circuit is thus of more importance than the material of which it is composed.

Von Bezold.—Von Bezold,* in 1870, discovered that electric currents or impulses are reflected from the insulated end of a conductor, and obtained nodes and loops of potential at different points on a wire. He was at the time experimenting with the well-known Lichtenberg dust figures. If a glass plate be covered on one side by tinfoil which is connected to earth, a discharge from the end of a conductor placed near the middle of the other side, leaves the bare surface of the glass electrified. Dust is then shaken over the surface and adheres to the glass in peculiar seaweed-like patterns, the forms of which for positive and for negative

* Poggendorf's *Annalen*, 140, p. 541, or "Electric Waves," Hertz, p. 54 (Macmillans). The following excerpts and diagrams, Figs. 7-13, are inserted by kind permission of Messrs Macmillan and Dr J. A. Barth, of Leipzic.

discharges are quite different in character. It is thus possible to distinguish a positive from a negative discharge though either may last only for a very minute fraction of a second.

The discovery of nodes on a conductor carrying oscillatory currents was made by means of the apparatus shown diagrammatically in Fig. 7.

By varying the lengths of the loops of wire D and D′, it was found that the dust figures produced at B disappeared, while those at A and C were strongly developed. A, B, and C were all equally near to the tinfoil earth-plate, but the result showed that the potential at B did not rise; while at A, which was earthed through a high resistance of great self-induction—what we would now call a choking coil—and at C, the completely insulated end, there existed loops of potential.

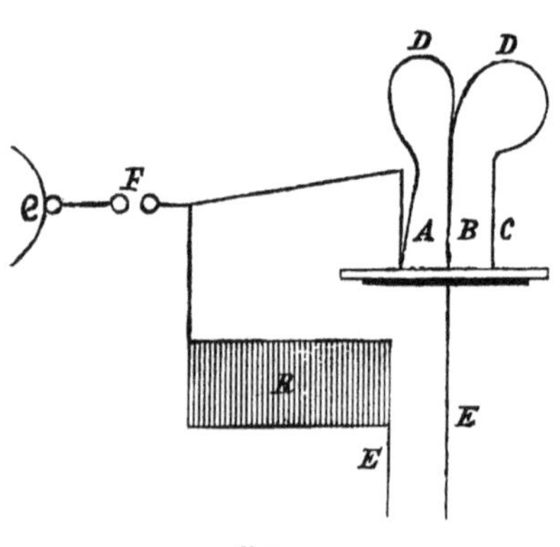

FIG. 7.

The reflection of electric waves from the insulated end of a conductor was thus proved by the production of stationary nodes and loops of potential and the existence of a definite velocity of propagation established.

The result of this research was probably the first experimental confirmation of Maxwell's theory of the propagation of electric disturbances, since it was now shown that their speed of propagation was finite.

Von Bezold sums up his results as follows:—

"1. If, after springing across a spark-gap, an electric discharge has before it two paths to earth, one short and the other long, and separated by a test plate, the discharge current splits up, so long as the sparking distance is small; but when it is larger the electricity rushes solely along the shorter path carrying with it, out of the other branch, electricity of the same sign.

"2. If a series of electric waves is sent along a wire which is insulated at the end, the waves are reflected at the end, and the phenomena which accompany this process in the case of alternating discharges appear to be caused by interference between the advancing and reflected waves.

"3. An electric discharge traverses wire of equal lengths in equal times whatever may be the material of which these wires consist."

It should be noticed that Von Bezold produced high frequency oscillations by the use of a spark-gap, a method which has not yet been superseded.

Hertz.—Eighteen years later Hertz published an account* of his experiments on the propagation of electric waves along wires. Though some of his conclusions have had to be modified in view of later knowledge, they are yet of such fundamental importance that we must summarise them :—

1. He shows that Faraday's view that electric forces are electric polarisations existing independently in space is confirmed by his experiments.

2. It is proved that the velocity of propagation of electric force in space is finite.

We now come to the series of experiments on which the fame of Hertz chiefly rests. As is usual in such cases, they did not originate in a sudden inspiration or happy accident, but were the continued development of the same theme by a mind endowed with wonderful insight, perseverance, and constructive imagination.

During the experiments described above he had often noticed that sparks appeared in the secondary circuit when

* Wiedemann's *Ann.* 34, p. 551.

it was in positions in which no direct action was possible, and in particular near walls or other bodies which might be supposed capable of reflecting the inductive action.

The simplest explanation appeared to be that the induction actually spread out as waves from the wire, which were then reflected from the walls, and, interfering with the advancing waves, caused regions corresponding to the nodes and loops in a stationary vibration. In order to find whether free radiation was in reality taking place the conditions of experiment were made more favourable for reflection and the whole space in the neighbourhood of the radiator thoroughly explored.

The radiator, which consisted, as in the last experiments, of two square brass plates, 40 by 40 cm., connected by a copper wire 60 cm. long, with a spark-gap at its middle point, was set up vertically at 13 m. distance from a wall on the face of which was a large sheet of zinc to act as reflector. The secondary conductor was, as before, a ring of wire of 35 cm. radius, which was mounted on a wooden stand capable of rotation about a vertical and a horizontal axis, and at one end of the horizontal axis there was a small adjustable spark-gap in the circle of wire. It was by the occurrence of minute sparks at this gap that the inductive action was observed, and the positions of the loops and nodes caused by the interference of the direct and reflected waves mapped out. At least this was the explanation which occurred to Hertz, and though it has been in some respects modified by the researches of Bjerknes and other experimenters, the main facts which he demonstrated, viz., (1) that true electro-magnetic waves, travelling as a free radiation through the dielectric may be generated by oscillatory currents on a conductor, and that (2) the velocity of these waves is finite, have been entirely confirmed by later researches.

Clerk-Maxwell had stated his belief in the existence of such waves, and had given, in part iv., chapter xx., of his treatise on "Electricity and Magnetism," the fundamental laws of their motion; but it was reserved for Hertz to give an experimental proof of their reality and to discover methods for their production and detection.

Hertz now proceeded to show that the waves produced by his oscillator were in all respects light waves of great length—that, in fact, they obeyed all the known laws of the propagation of light. This was not accomplished with the original oscillator, since the waves it produced were several metres long, rendering necessary the use of excessively large reflectors and prisms to deal with them. A much smaller oscillator was therefore made which gave out a wave of about 60 cm. This apparatus consisted of two thick brass rods with a short spark-gap between them, placed along the focal axis of a cylindrical parabolic mirror of sheet metal. The receiver was somewhat similar in shape, and was also placed in the focal line of a mirror. It had an adjustable spark-gap for detection of the induced oscillations.

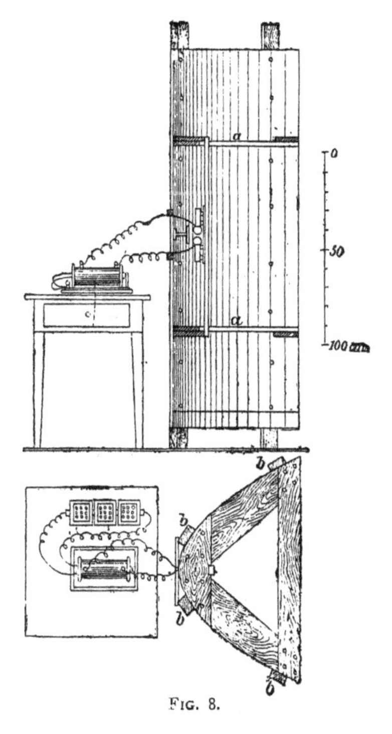

Fig. 8.

With this apparatus it was possible to demonstrate, by using a flat sheet of metal 2 m. square, that the waves followed the law of reflection, *i.e.*, that their angle of reflection from the mirror was equal to their angle of incidence. It was also shown that if the axis of the receiving mirror were horizontal while the transmitter was vertical, no sparks were visible in the former, thus proving, as was to be expected, that the waves emitted were plane polarised. This conclusion was confirmed by placing a grid of parallel copper wires in the path of the ray. Both transmitter and receiver being plated with their axes vertical, if the wires of the grid were also vertical it formed a perfect screen, nothing being observable at the receiver; but if the wires were horizontal, *i.e.*, perpendicular to the axes of the mirrors, transmission was practically as good as if there were no grid there.

In the first position it was found by experiment that the action of the grid was not merely absorption, but that

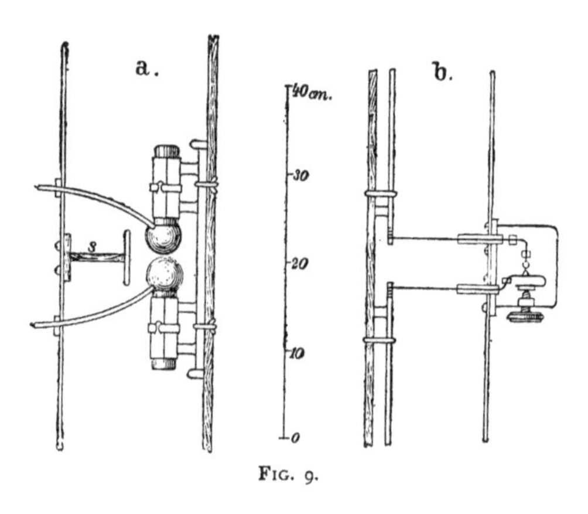

FIG. 9.

a large proportion of the radiation was reflected from it as from a plane mirror. In this respect the phenomenon differs from the action of a tourmaline plate on light, since the tourmaline absorbs what it does not transmit.

A large prism of pitch, having as a base an isosceles triangle of 1.2 m. in the side and whose height was 1.5 m., was next constructed, and was used to determine whether the rays were refracted on passing from air into a medium of different refrangibility. The angle of the prism was about 30°. It was found that the ray deviated about 22° on entering or leaving the prism; as the angle of the prism was 30° the refractive index of the pitch was therefore 1.69, a value which is within 5 per cent. of its refractive index for ordinary light.

Thus Hertz showed that free electric waves exist, and that light, as Clerk-Maxwell had foretold, is merely a manifestation of them.

The following is from Hertz's description of the form and propagation of the electric waves :*—

"We shall content ourselves with considering the results of the construction as shown in Figs. 10, 11, 12, and 13. These figures exhibit the distribution of force at the times, $t=O$, $\frac{1}{4}T$, $\frac{1}{2}T$, $\frac{3}{4}T$, or by a suitable reversion of the arrows for all subsequent times which are whole multiples of $\frac{1}{4}T$. At the origin is shown, in its correct position and approximately to correct scale, the arrangement which was used in our earlier experiments for exciting the oscillations. The lines of force are not continued right up to this picture, for our formulæ assume that the oscillator is infinitely short, and therefore became inadequate in the neighbourhood of the finite oscillator.

"Let us begin our explanation of the diagrams with Fig. 10. Here $t=O$; the current is at its maximum strength, but the poles of the rectilinear oscillator are not charged with electricity—no lines of force converge towards them. But from the time $t=O$ onwards, such lines of force begin to shoot out from the poles; they are comprised within a sphere represented by the value $Q=O$. In Fig. 10, indeed, this sphere is still vanishingly small, but it rapidly enlarges, and by the time $t=\frac{1}{4}T$ (Fig. 11) it already fills the space R. The distribution of the lines of force within the sphere is nearly of the same kind as that corresponding to a static electric charge upon the poles. The velocity with which the spherical surface $Q=O$ spreads out from the origin is at first much greater than $1/A$; in fact, for the time $\frac{1}{4}T$, this latter velocity would only correspond to the value of $\frac{1}{4}$ given in the figure. At an infinitesimal distance from the origin the velocity of propagation is even infinite.

---

* "Electric Waves," Hertz, p. 146, translated by D. E. Jones (Macmillan). Inserted by kind permission of Messrs Macmillan and Dr J. A. Barth.

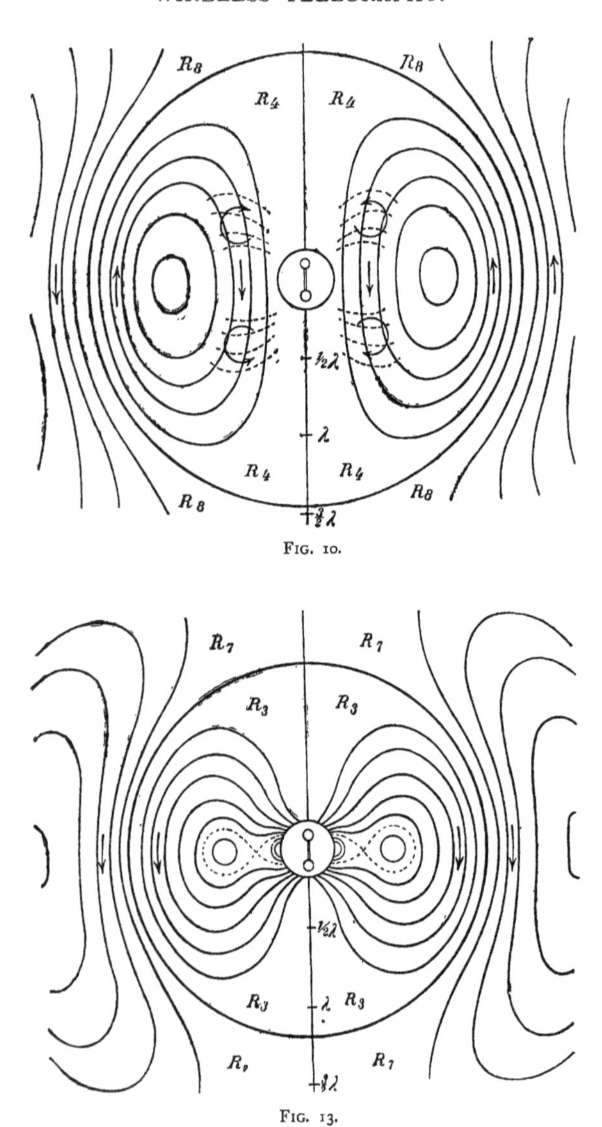

Fig. 10.

Fig. 13.

"This is the phenomenon which, according to the old mode of expression, is represented by the statement that

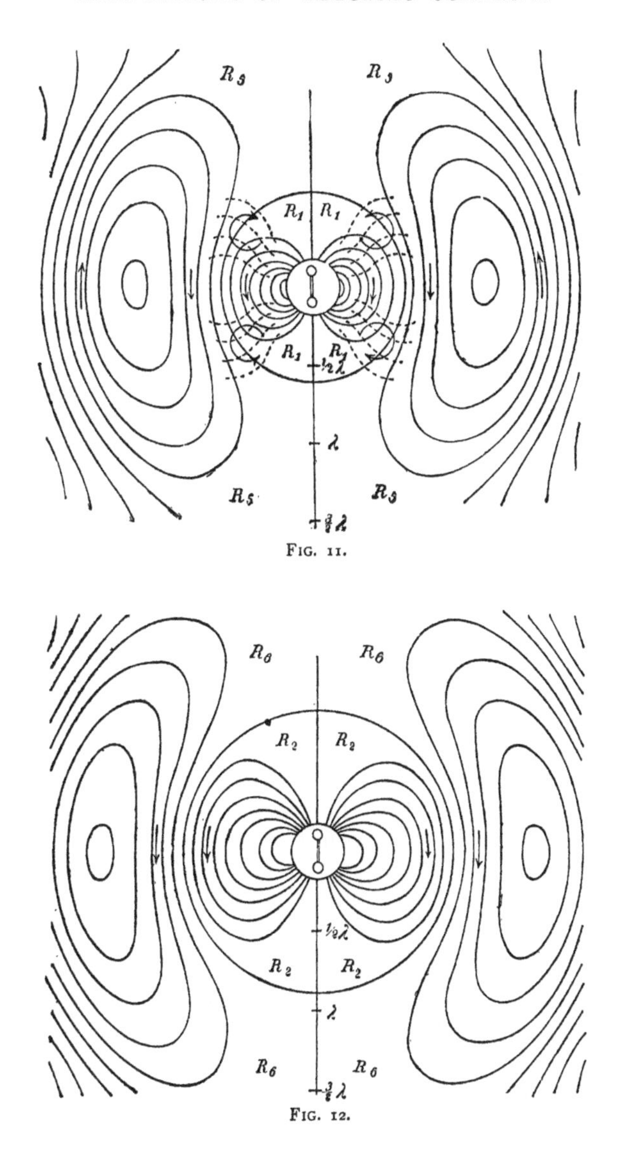

FIG. 11.

FIG. 12.

upon the electro-magnetic action which travels with the velocity $1/A$, there is superposed an electrostatic force travelling with infinite velocity. In the sense of our theory we must correctly represent the phenomenon by saying that fundamentally the waves which are being developed do not owe their formation solely to the processes at the origin, but arise out of the conditions of the whole surrounding space, which latter, according to our theory, is the true seat of the energy. However this may be, the surface $Q=O$ spreads out further with a velocity which gradually sinks to $1/A$, and by the time $t=\frac{1}{2}T$ (Fig. 12) fills the space R_2. At this time the electrostatic charge of the poles is at its greatest development; the number of lines of force which converge towards the poles is a maximum. As time progresses further no fresh lines of force proceed from the poles, but the existing ones rather begin to retreat towards the oscillating conductor, to disappear there as lines of electric force, but converting their energy into magnetic energy. Here there arises a peculiar action which can plainly be recognised, at any rate in its beginnings, in Fig. 13 ($t=\frac{3}{4}T$). The lines of force which have withdrawn furthest from the origin become laterally inflected by reason of their tendency to contract together; as this inflection contracts nearer and nearer to the z-axis, a portion of each of the outer lines of force detaches itself as a self-closed line of force which advances independently into space, while the remainder of the lines of force sink back into the oscillating conductor.

"The number of receding lines of force is just as great as the number which proceeded outwards, but their energy is necessarily diminished by the energy of the parts detached. This loss of energy corresponds to the radiation into space. In consequence of this, the oscillation would of necessity soon come to rest unless impressed forces restored the lost energy at the origin. In treating the oscillation as undamped, we have tacitly assumed the presence of such forces. In Fig. 10 to which we now return for the time $t=T$, conceiving the arrows to be reversed—the detached portions of the lines of force fill the spherical space R_4, while the lines of force proceeding from the poles have completely disappeared. But new lines of force burst out from the poles and crowd together the lines whose development we have followed into the space R_5 (Fig. 11). It is not necessary to explain further how these lines of force make their way to the spaces R_6 (Fig. 12), R_7 (Fig. 13), R_8 (Fig. 10).

They run more and more into a pure transverse wave-motion, and as such lose themselves in the distance. The best way of picturing the play of the forces would be by making drawings for still shorter intervals of time and attaching these to a stroboscopic disc.

"A closer examination of the diagrams show that at points which do not lie either on the z-axis or in the xy-plane the direction of the force changes from instant to instant. Thus, if we represent the force at such a point in the usual manner by a line drawn from the point, the end of this line will not simply move backwards and forwards along a straight line during an oscillation, but will describe an ellipse.

"In order to find out whether there are any points at which this ellipse approximates to a circle, and in which, therefore, the force turns successively through all points of the compass without any appreciable change of magnitude, we superpose two of the diagrams which correspond to times differing by $\frac{1}{4}$T from one another, *e.g.*, Fig. 10 and Fig. 12, or 11 and 13. At such points as we are trying to find, the lines of the one system must clearly cut those of the second system at right angles, and the distances between the lines of the one system must be equal to those of the second. The small quadrilaterals formed by the intersection of both systems must therefore be squares at the points sought. Now, in fact, regions of this kind can be observed; in Figs. 10 and 11 they are indicated by circular arrows, the directions of which at the same time give the direction of rotation of force. For further explanation dotted lines are introduced which belong to the system of lines in Figs. 12 and 13. Furthermore we find that the behaviour here sketched is exhibited by the force not only at the points referred to, but also in the whole strip-shaped tract which, spreading out from these points, forms the neighbourhood of the z-axis. Yet the force diminishes in magnitude so rapidly in this direction that its peculiar behaviour only attracts attention at the points mentioned."

From these researches has sprung modern wireless telegraphy, and though free radiations are no longer generally used, Marconi, among others, drew his original inspiration from Hertz, his first essay at wireless telegraphy being the invention of a telegraphic receiver which could detect and record as Morse signals the arrival of waves from a Hertz transmitter of the form we have described.

In 1894, at a lecture in the Royal Institution, Sir Oliver Lodge showed how the coherer might be utilised for the detection of Hertzian waves up to a distance of about 150 yards. The receiver, however, could hardly be called a telegraph instrument, and the possibility of developing it into one was apparently at the time hardly realised by the

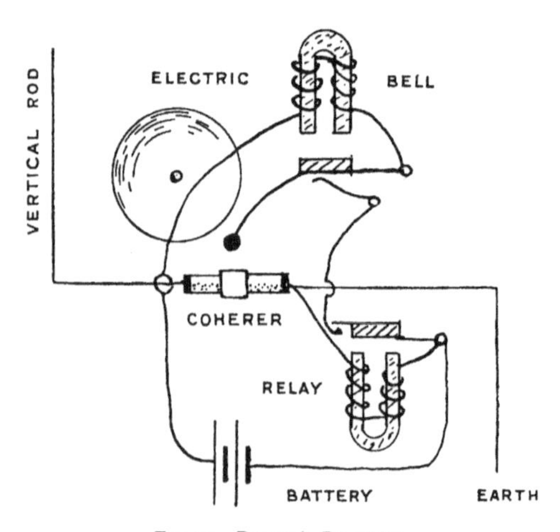

FIG. 14.—POPOFF'S RECEIVER.

lecturer. His syntonic system, patented in 1897, will be described later.

Early in 1895 Professor Popoff of St Petersburg described a system of observing atmospheric electricity by means of a lightning rod connected through a coherer to earth. A relay and local battery were put in circuit with the coherer, and the relay in turn closed the circuit of an electric bell. This arrangement, which resembles Marconi's original receiver, the inventor proposed in December 1895 to use for telegraphic purposes, if a sufficiently powerful generator of electric waves could be found to serve as transmitter. It was apparently not until after Marconi's success had become known that Professor Popoff discovered the necessary transmitter. Fig. 14 shows, diagrammatically, the arrangement for detecting electrical disturbances which it was proposed to use as a telegraphic receiver. The system is now in use in France and Russia.

Nikola Tesla.—It is impossible to leave the part of our subject which is concerned with the fundamental discoveries on which wireless telegraphy is based without taking account of some of the brilliant inventions and schemes of Tesla. Though much that has been attributed to him has undoubtedly had its origin in the prolific imagination of the American Press, there remains an exceedingly substantial basis of fact, mostly in the form of patent specifications, or of actual lecture demonstrations, which shows his extraordinary inventive power and wonderful knowledge of the properties of alternating currents of both low and high frequency.

Among many other inventions, made as early as 1893, perhaps the most important to wireless telegraphists is his method of producing long trains of waves of high frequency, and of transforming them to higher voltage. After several unsuccessful attempts he completed an alternator which could be run at 30,000 periods per second, and designed a form of transformer capable of transforming these currents to very high voltage. He also showed that his transformer, or "Tesla coil," as it is usually called nowadays, could transform currents of much higher frequencies than were obtainable from his alternator, even currents of 100,000 or 1,000,000 periods per second, such as are produced by the oscillatory discharge of a Leyden jar. Fig. 15 shows the apparatus used.

Even at low voltages high frequency currents show many remarkable peculiarities, most of which are not noticeable in ordinary alternate current working, unless in very exceptional circumstances. The heating effect of a high frequency current is, for instance, enormous owing to its limitation to a thin skin on the surface of the conductor. The hysteresis losses in iron placed in so rapidly alternating a field are also very great. Thus Tesla found that the temperature of an iron wire $\frac{1}{16}$ inch in diameter inside a coil of 250 turns, in which a current estimated at 5 amperes was flowing, rose about 500° C. in two seconds.

At very high frequencies the presence of an iron core in a coil did not appreciably add to its self-induction, an interesting fact of immediate application in wireless telegraphy.

Remarkable effects may be produced by adjusting the capacity and inductance of a high frequency circuit. Thus, Tesla found that if a condenser be included in the circuit, the potential at its terminals may rise far above that of the alternator. That similar effects are not in any appreciable degree noticeable with lower frequencies is probably due to the enormous condensers that would be required in order to obtain electrical resonance (revibration) at the low frequencies which are used in the distribution of electrical energy. To determine the actual size of condenser which would be required in this case, let us assume that an inductance such as that of the field magnets of a dynamo is available, and then find by calculation the capacity which must be added to give a natural frequency of 50 periods per second to the circuit.

The natural frequency of an electric vibration in a circuit of capacity C, microfarads, and inductance, L, cms., is given by the equation—

$$n = \frac{5{,}000{,}000}{\sqrt{CL}}$$

If we take L = 1,000,000 cm., and $n = 50$, we find C approximately equal to 10,000 microfarads. A condenser of this capacity, if constructed on the usual plan, would occupy about 10 cubic metres. If exactly adjusted to equifrequency with the alternator, the circuit would exhibit resonance effects on an extraordinary scale.

Let us return, however, to Tesla's experiments, the results of which throw a flood of light on many points in wireless telegraph working which would otherwise be hard to understand. When a high frequency current is passed through the primary of a step up transformer the voltage is greatly increased, and many new phenomena are observable. Powerful brush and flame discharges are given off by conductors connected to the terminals, the neighbouring air being actually rendered quite warm by the intense molecular bombardment, even though the frequency be not above 10,000 per second. With a frequency ten times as great as this the effects are much more marked; to obtain them, however, an alternator is not sufficient, the greatest frequency Tesla obtained by such means being about 30,000 per second. An induction coil was therefore used, which charged a Leyden jar. The latter discharged itself across a spark-gap through the primary of a Tesla transformer. The discharge of the jar being oscillatory, the currents induced in the secondary of the transformer were of much higher voltage, but of the same frequency. At so great a frequency the impedance of even a thick short conductor is so large, that differences of potential amounting to hundreds of volts may be maintained at points quite near together; nodes and loops of potential, which may be located by means of an ordinary glow lamp, also appear on the conductor.

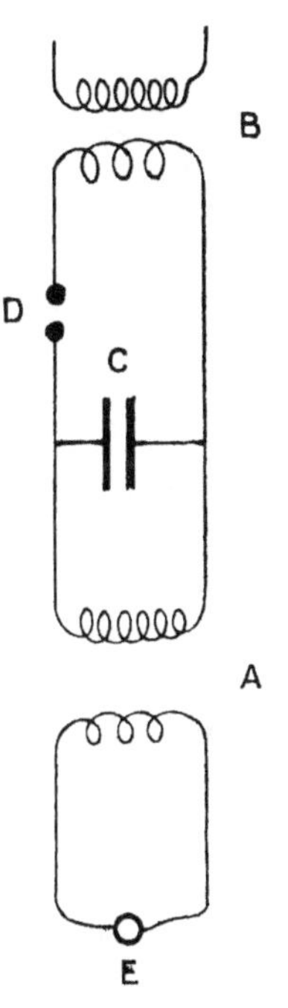

FIG. 15.—TESLA TRANSFORMER FOR PRODUCING HIGH FREQUENCY ALTERNATING CURRENTS OF HIGH VOLTAGE.

A, Transformer; *B*, Special Transformer; *C*, Condenser; *D*, Spark-Gap; *E*, Alternator.

The transformer used to produce these effects was of

necessity very different from the ordinary article of commerce. The primary consisted of a few turns of thick copper wire inside a glass tube. No iron core was used, as with such high frequencies the magnetic induction does not penetrate the iron, and thus no advantage would be

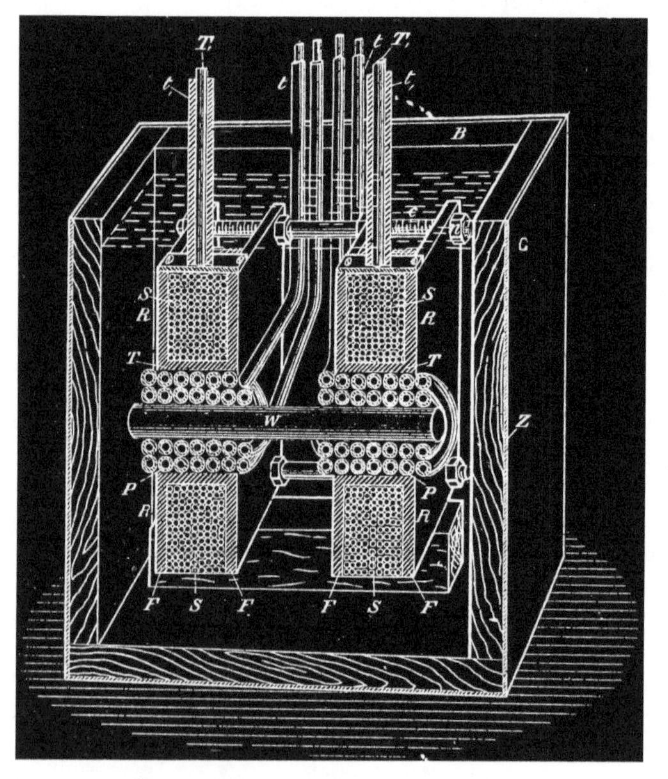

FIG. 16.—TESLA'S DISRUPTIVE DISCHARGE COIL.
P, Primary of G.P. covered wire; *S*, Secondary; *R R*, Bobbins containing Secondary.

gained. Outside the glass tube were two small bobbins, wound with guttapercha-covered copper wire. The number of turns was not great, as the rise of voltage depends mainly on resonance in the secondary circuit, and not directly on the ratio of the numbers of turns in the primary and secondary. The whole transformer was sunk in a bath of insulating oil from which air had been removed by means of a pump. In order to get the maximum effect the frequency of the secondary circuit is adjusted to that of the primary by altering its capacity. This may be done by means of a small condenser, or by merely altering the size of the spheres which form the secondary terminals. Tesla has shown that in dealing with these currents, liquid dielectrics are far preferable to solids as insulators. The molecular bombardment heats the surface of a solid, or the gas contained in any cavity of it, so rapidly, that it soon loses its insulating properties. Oil, free from air, gives three or four times as good results as the best solid insulator, and has the advantage that if a disruptive discharge should occur through it, the puncture is self-sealing.

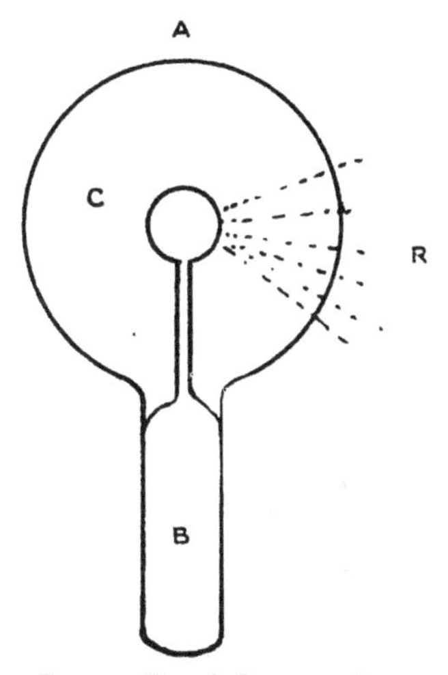

FIG. 17.—TESLA'S SENSITIVE RAY.
A, Outside Glass Bulb (High Vacuum); *B*, Inner Bulb (Low Vacuum), connected by tube with *C*; *R*, Sensitive Rays.

A transformer circuit of this type is in use in almost all wireless telegraph stations for the production of the electric waves of high frequency and voltage which are used in the transmission of signals.

One very curious phenomenon—which Tesla suggests may be useful in wireless telegraphy, but which (as far as the writer is aware) has not yet been so employed—remains to be mentioned. It may be called the sensitive ray. A small spherical glass bulb, on a tubular stem, is sealed into a larger spherical bulb. The space between the bulbs is very highly exhausted, and the air in the small bulb and stem is rarified just sufficiently to make it conductive. If one terminal of the Tesla transformer is connected to the air in the stem, a soft light appears round the central sphere. In time this diffuse radiation condenses into a definite ray which is extremely sensitive to both electric and magnetic influence. A small permanent magnet only 2 cm. in length will affect the ray at a distance of several metres, and the mere presence of a conducting body in the neighbourhood of the bulb causes the ray to move so that it is always turned away from the conductor. Unfortunately this condition is not permanent, though when the phenomenon is properly understood means may be found to make it so. The discovery was made fully a dozen years ago, before wireless telegraphy had become practical; perhaps it may some day be revived and form part of a peculiarly sensitive receiver.

The Superposition of Small Periodic Motions.—A favourite lecture experiment of Lord Kelvin's is to show the gradual increase in the amplitude of the swing of a long and heavy pendulum when its bob is struck with a very small mass, at the times when it reaches the limit of each swing. A factory has been known to collapse through the motion of the engine getting into time with the natural period of vibration of the structure. All are illustrations of a general principle of immense importance in nature, and of paramount importance in the working of a wireless system.

In wireless telegraphy the two great desiderata are the ability to transmit messages to great distances, and to prevent interference between neighbouring stations. The solution of both problems has, so far, depended mainly on the principle of the superposition of small motions. In order that this principle may be successively applied it is necessary that the oscillations should continue, with little diminution of amplitude for many complete periods, *i.e.*, that the damping should be slight. Damping or decrease of amplitude is due to two causes—firstly, to dissipation of energy, as heat, in the local circuits; secondly, to radiation. The first may be reduced to a minimum by properly proportioning the resistance, capacity, and inductance of the circuits, but the second is inevitable, as it represents the energy transmitted outwards, a portion of which actuates the distant receiver.

Now the action of a receiver usually depends on the

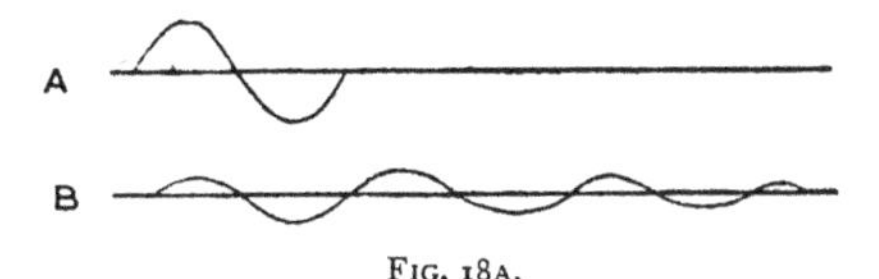

FIG. 18A.
A, Transmitted Wave; *B*, Oscillatory Current produced in Receiver.

amplitude of the oscillation of potential excited in it by the waves from the transmitting station. If these consist, as in Marconi's untuned system of one wave of large amplitude with a train of three or four rapidly diminishing waves behind it, it is clear that the reception of signals depends entirely on the amplitude of the first wave and that the energy of the others is quite negligible in comparison. If,

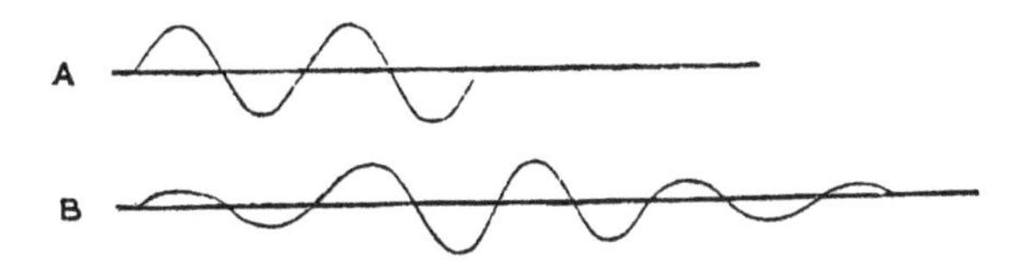

FIG. 18B.
A, Two Waves Transmitted; *B*, Resultant Current in Receiver, showing growth through superposition of impulses.

however, a long train of very slowly diminishing waves be sent out, the effect of each, if the receiver be properly constructed, will be added to that of those before until an oscillation of great amplitude is induced in the receiver. It is more easy to transmit a large amount of energy by a succession of small waves than by one large one, for the latter would involve excessively high voltage and current. Hence the advantage, in long distance transmission, of using transmitters and receivers whose natural frequencies are equal. Thus if the frequencies of the electrical vibrators at both stations be the same and the local dissipation of energy very small, it is possible to communicate by using long trains of waves of moderate voltage. Distances may be covered in this manner which would be impracticable if transmission depended on a single wave or impulse.

It is fortunate that the best conditions for the transmission of signals are also those which render it possible to arrange for the non-interference of neighbouring stations. If a receiver only acts after a considerable number of properly timed waves have arrived, we may arrange that it is not actuated by other transmitters by causing these to give out waves of frequencies other than that to which the receiver responds. The weak point of this system of discrimination between stations is that the receiver may respond to trains of waves of slightly greater or less frequency than its own. A certain margin must therefore be allowed for each station in the matter of frequency, and therefore the number of non-interfering stations in a neighbourhood is not unlimited. This question has been recently discussed by J. J. Hettinger,* who has determined the limits of non-interference of two stations.

If a piece of elastic material, such as a metal rod, or the column of air in an organ pipe, be caused to vibrate, it has been found that the motion tends to take certain definite forms. Thus a stretched string will vibrate as a whole, the only parts not in motion being the fixed ends. This is called its fundamental vibration. Or it may have a quiet point at its middle while the rest of its length is in motion. This mode of vibration is called the first harmonic, or octave (or sometimes the second harmonic). Its frequency is double that of the fundamental vibration. In like manner the vibrating body may move in 3, 4, 5, or any number of

---

* *Electrical Review*, 1906.

equal segments; the corresponding frequencies of the emitted waves being 3, 4, 5, or n times that of the fundamental vibration. The motion of the electricity on an electrical vibrator, by which is meant a conductor, or combination of conductors, bounded by a dielectric which reflects the electrical surges back on themselves from the ends of the conductor and so maintains a "stationary" vibration of definite period, follows a similar law, the frequency of the vibration being proportional to the number of segments in which it takes place.

Electrical engineers have recently had this subject of harmonic vibration brought very clearly before them through the study of alternate currents by means of Duddell's Oscillograph; and though the harmonics in this case are as a rule forced and not due to natural resonance, the compound

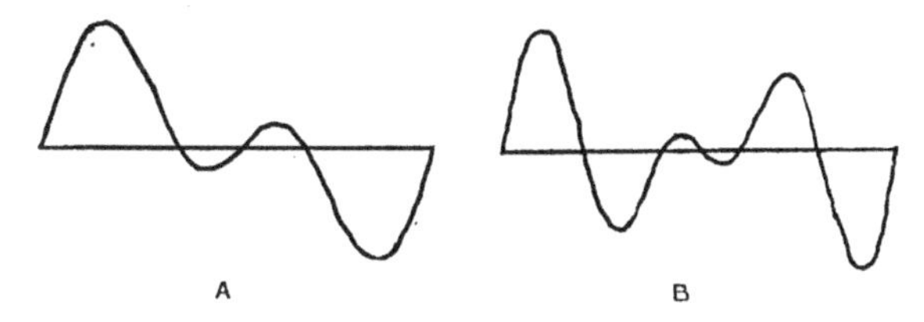

FIG. 19.—COMPLEX PERIODIC VIBRATIONS.
A, Fundamental and First Overtone (Frequencies, 1 : 2); *B*, Fundamental and Second Overtone (Frequencies, 1 : 3).

waves obtained are of the same character as those formed by the synthesis of natural harmonics. A few types of compound waves are shown in Fig. 19.

The various phenomena connected with the production of stationary electric waves on wires have been very fully elucidated by a number of experimenters. Among many others of lesser importance, the names of Von Bezold, Lodge, Hertz, Sarasin and De la Rive, and Barton* may be mentioned. Their researches have proved that the rate of propagation of waves of high frequency along wires is practically

* *Phil. Mag.* May 1899.

the same as the velocity of light, *i.e.*, about 3×10^{10} cm. per second. In wireless telegraphy we have to deal with waves which travel outwards from a centre along a nearly plane conductor as well as with waves along wires; the theory of the former is not so well understood, most of our knowledge of the subject is therefore rather qualitative than quantitative. In long-distance transmission we have also to take into account the conductivity of the rarified air of the upper atmosphere, which may reduce the propagation of the waves to the case of two concentric spherical conductors. These questions will be considered in a later chapter.

In 1896 Professor E. Rutherford, using a large horizontal Hertzian radiator as transmitter, and his magnetic detector with short unearthed conductors as receiver, obtained signals at distances of over a mile.

CHAPTER II.

EARLIER ATTEMPTS AT WIRELESS TELEGRAPHY.

HAVING briefly described in the last chapter the experimental researches which showed the way to the invention of modern high frequency wireless telegraphy, we must now go into the actual experiments, directed definitely to the invention of a new means of communication, from which the systems now in use have sprung.

Hughes's discovery in 1879, that his microphone could detect, at a considerable distance, the oscillations radiated from a conductor connected to one side of the spark-gap of an induction coil, came before the time was ripe for such things, and as they were not published, cannot have influenced subsequent workers. They stand as a warning to the inventor against discouragement by the dicta of scientific men, who, though learned, have not gone through in detail the course of experiment and reasoning which have led him to his conclusions.

The coherer, subsequently rediscovered by Calzecchi-Onesti, Lodge, and Branly, was simply an imperfect contact like the microphone, and if Hughes's results had been published in 1879 there is little doubt that the existence of electrical radiations would have been an established fact several years before Hertz's experiments were carried out.

Probably the earliest successful attempt at communication by means of a system in which only one earth connection is used at each station was that of Professor Dolbear, who in 1882 applied for an American patent for a system of electrical communication between two or more places with-

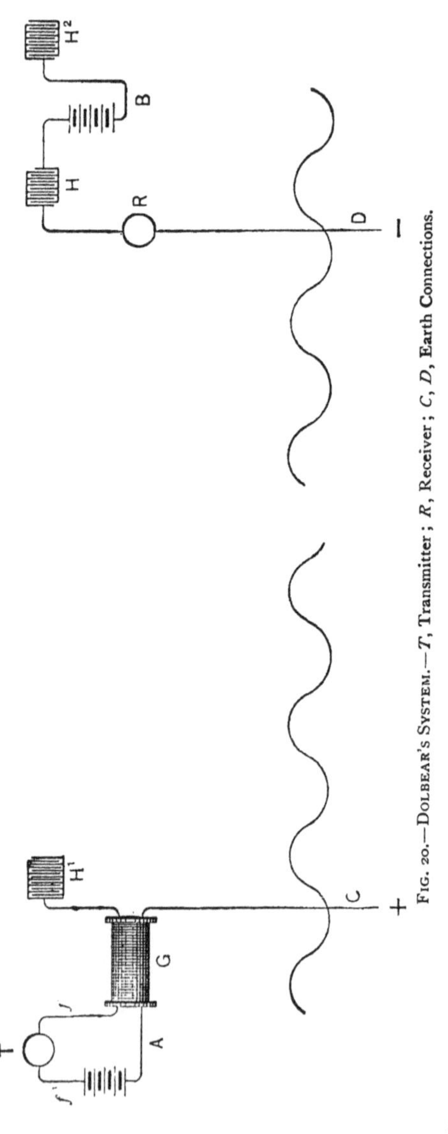

Fig. 20.—Dolbear's System.—*T*, Transmitter; *R*, Receiver; *C*, *D*, Earth Connections.

out wire or conductor. In his first experiment he employed a small magneto-electric machine, *i.e.*, a producer of alternating current, as transmitter. One terminal only was earthed, the other being free and only a foot or two long.

The receiver was an ordinary telephone, one terminal being to earth and the other held in the hand of the observer, who was insulated. The distance of transmission in the first instance was about sixty feet, which was later increased by substituting an induction coil, with a Morse key in the primary circuit, for the magneto; large capacities were also attached to the free ends of the receiver and transmitter, and apparently even telephonic speech was transmitted over a distance of about half a mile. The patent specification does not greatly elucidate the theory of the system. The author states that it is of prime importance to keep the earth wires oppositely charged at transmitter and receiver, and for this purpose he attaches the positive and negative poles of primary batteries to them. What advantage this arrangement could give, when the transmission was in reality by *alternate* currents, is not at all clear. Fig. 20 shows the apparatus used on this system when intended for the transmission of speech.

Still following the chronological order, we find that in 1893 Tesla proposed a plan for the wireless transmission of electrical energy—including, of course, telegraphic messages—which, though resembling Dolbear's system in

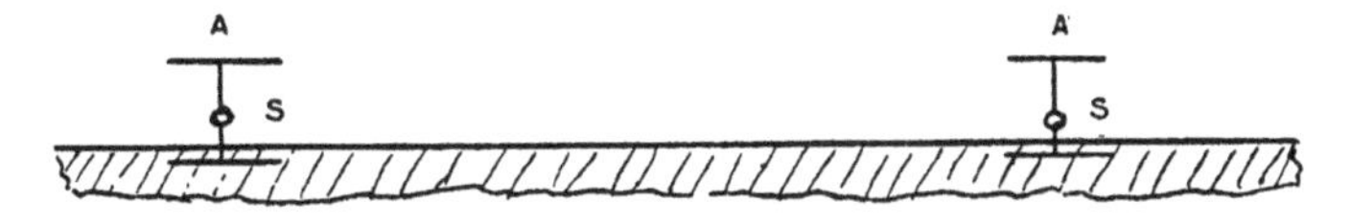

FIG. 21.—TESLA'S PROPOSED WIRELESS TELEGRAPH.
A, *A'*, Aerial Capacities ; *S*, *S'*, Transmitting and receiving instruments.

some respects, contained an important, indeed a fundamental improvement.

The proposal was to use an elevated capacity and a single "earth" at each station. At the transmitter a source of alternating currents was to be employed as in Dolbear's transmitter, and at the receiver some type of instrument suitable to the frequency of the current employed. The improvement lay in the suggestion that the self-induction and capacity of the receiver should be adjusted so that its natural period of electrical vibration should be the same as that of the transmitted current. The receiver would thus revibrate to the waves received, and the oscillations generated in it grow in amplitude as they continued to arrive. A series of very slight waves of low potential would therefore be able to create, in time, an oscillation great enough to be recorded by the instrument. This suggestion grew naturally out of the inventor's discoveries of the properties of high frequency currents, but was apparently not put into practice, nor even patented, until 1897. Earlier in the same year, Sir Oliver Lodge, approaching the subject from the point of view of Hertzian radiations, had patented a system of telegraphy without wires in which Hertzian waves were employed. The transmitter was a species of Hertz oscillator, the receiver being of the same electrical dimensions (*i.e.*, capacity, &c.), and having therefore the same natural rate of vibration as the transmitter. This is the first patent extant for a syntonic system. Tesla's patent, based on his suggestion of 1893, was taken out somewhat later. It should be noted that he did not, like Lodge, propose to employ free radiations, but dealt with oscillating currents conducted by the earth.

Captain H. B. Jackson, R.N., in 1895, before Marconi's inventions had been published, had actually succeeded in transmitting in the Morse code between two ships by means of apparatus very similar to that simultaneously invented by Marconi. Details of Captain Jackson's apparatus have never been published, as the system is of course the property of the Navy. Professor Trouton had also experimented on a similar plan.

Marconi.—As already mentioned in the last chapter, Mr Marconi's first wireless system (patented in 1896) was based directly on Hertz's experiments. The transmitter consisted of a Hertzian oscillator placed in the focal line of a parabolic mirror and actuated by an induction coil. The receiver was also similar to that of Hertz, with the addition of a coherer as detector and an automatic arrangement for causing decoherence immediately on reception of a signal. These additions converted an apparatus which had been merely a detector of electrical waves into a telegraphic instrument which could transmit messages by their agency. Using direct radiations, Marconi succeeded with this apparatus in communicating over a distance of about 2 km., a distance at which no previous experimenter had been able even to detect the arrival of the waves.

The Righi form of oscillator was also used in many of the experiments. It differs from Hertz's in having two isolated metal spheres between the terminal spheres of the high tension circuit of the induction coil. There are thus three spark-gaps in series, two long ones at either end and a short one in the middle between the insulated spheres. It gives out waves of very high frequency, as its period is determined by the dimensions of the two spheres.

Marconi had not continued his experiments long with this apparatus when it occurred to him to increase its transmitting power by connecting large insulated conductors to each side of the spark-gap. Similar conductors were attached to the receiver circuit, and when finally the earth was requisitioned as one conductor, while the other was raised high above it on a pole, it was found possible to read the messages at eight or ten times the distance at which they could be received by Hertzian radiation alone.

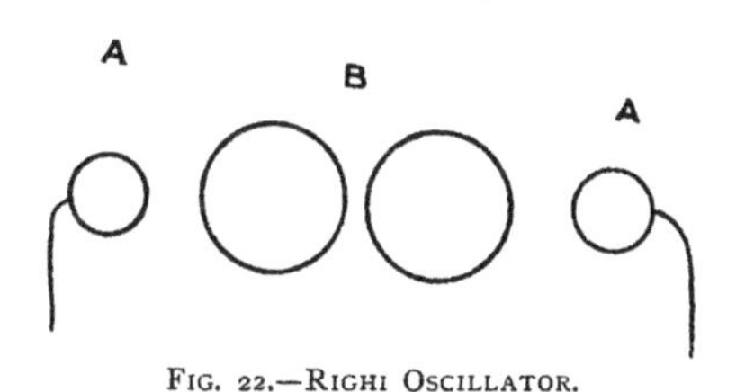

Fig. 22.—Righi Oscillator.
A, *A'*, Small Spheres connected to source of current; *B*, Large Spheres constituting Main Oscillator.

The practical value of this vast improvement was at once appreciated, but the complete change which had been effected in the method of transmission by making the earth connection was by no means so quickly recognised. It seems as if the glamour of Hertz's discovery of free electric radiation had blinded the majority of physicists to the otherwise obvious fact that the elevated capacity and the earth had now become the oscillator, and that the receiver was situated on a remote part of the oscillator itself. It is curious that the conducting power of the earth should have been overlooked after it had served for so many years as the return conductor for telegraphic and telephonic currents.

Marconi's first patent includes (1) direct Hertzian radiation apparatus, and (2) oscillatory current apparatus with one terminal elevated and one to earth at each station. This latter, he remarks, is better for communicating "through or across hills and other obstacles." Figs. 23 and 24 show the general arrangement of transmitter and receiver in this case. It will be noticed that the Righi oscillator was adhered to in the transmitter, though, in the light of later knowledge, it apparently served no

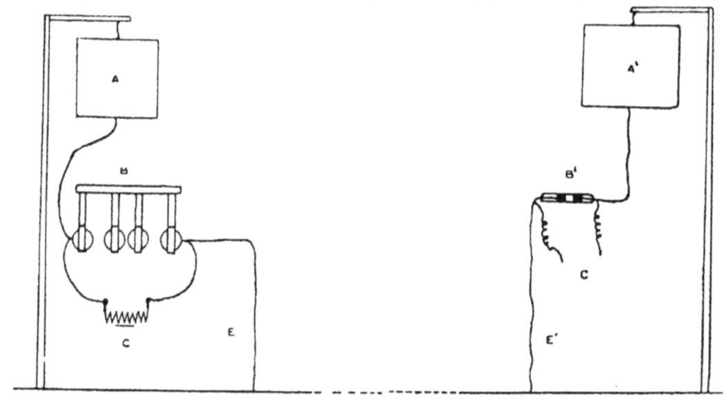

FIG. 23.—MARCONI'S EARTHED SYSTEM (Early Form).
A, *A'*, Elevated Capacities; *B*, Righi Oscillator; *C*, Source of Current (Induction Coil); *E*, *E'*, Earth Connections; *B'*, Coherer; *C'*, Receiver Connections.

useful purpose. The principal oscillator was now formed by the aerial capacity and the earth, and gave out oscillations of much lower frequency than the Righi spheres. The whole function of the latter was therefore to convert a little of the available energy into free radiations of very high frequency, which at very moderate distances became so dispersed as to be of no practical use in transmission. Within the course of a year, Mr Marconi, taught by experience, discarded the Righi oscillator in favour of a single spark-gap.

Marconi's first receiver merits description, as though

the receivers now in use contain very many new features, it was a thoroughly practical telegraphic instrument, and if in its early form it had less reliability than an ordinary telegraphic instrument, it could provide a means of communication in circumstances in which the latter would be totally useless.

The terminals of the coherer were connected to the wire from the otherwise insulated aerial capacity and to the earth wire. One end of the coherer was also connected to a dry cell, and the other through the magnet coil of a

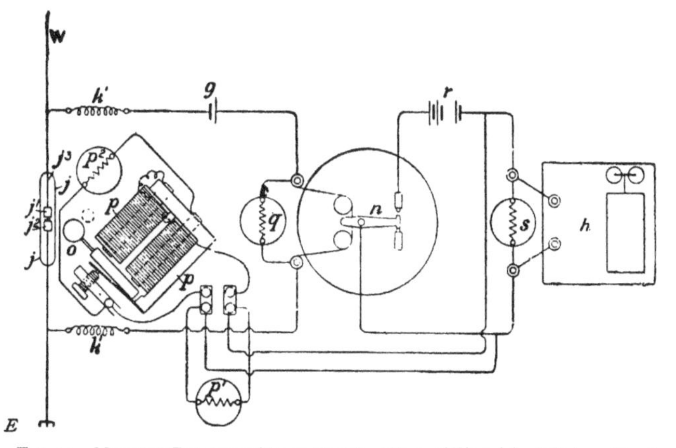

FIG. 24.—MARCONI RECEIVER (Early Form). [From "Electrician" Primer No. 67, by kind permission of the proprietors.]

W, Aerial Wire; *E*, Earth Wire; *j*, Coherer; *o*, Tapper; *k*, *k'*, Choking Coils; *g*, Dry Cell; *n*, Relay; *r*, Battery; *h*, Morse Inkwriter; p^1, p^2, *q*, *s*, Shunts.

relay to the other terminal of the cell. In the outer circuit of the relay a battery of eight cells and an electric bell, in which the coherer took the place of the gong, were put in series. Small choking coils were inserted between the coherer and the relay circuit to prevent the oscillatory currents being dissipated in the relay, and shunts were put in parallel with all parts of the circuit in which oscillations might be generated. In parallel with the tapper circuit was a Morse inkwriter or other recording instrument. The whole receiver, except the inkwriter, was enclosed in a metal box to screen it from the action of the transmitter in the station. The box was connected to earth, the aerial wire being temporarily introduced during the reception of signals through a hole in one side. The unearthed terminal of the inkwriter was connected to the apparatus inside the box by an ingenious arrangement, which, while transmitting uni-directional currents, did not admit oscillatory currents. A length of guttapercha-covered wire was coated with tinfoil, and after being made into a coil was placed in electric contact with the outside of the box. One end of this wire was led inside, and the other connected to the inker. Direct currents, therefore, could pass along the wire, but oscillatory currents preferred to pass across from the wire to the tinfoil coating as dielectric currents, and thence *via* the surface of the box to earth. This arrangement made it unnecessary to disconnect the inker while transmitting signals from the station.

With this apparatus messages were transmitted about forty miles over Salisbury Plain, kites being used to elevate the aerial wire. A demonstration was given with the assistance of the Post Office across the Bristol Channel at Lavernock, where it is about seven miles wide, the aerials being supported by poles 150 feet high.

Experimental stations were then put up at the Needles and Bournemouth, their distance apart being about fourteen miles; the latter station was removed, in October 1898, to the mouth of Poole Harbour, a distance of sixteen miles from the Needles. Temporary installations were also put up on board the royal yacht, for communication with Ryde, and on a tug boat off Kingstown to report the progress of the yachts during a regatta.

The experimental development of the system may be followed by reference to the records of the Patent Office and to the reports of Mr Marconi's occasional lectures to scientific societies. Early in the course of the experiments the large elevated capacity was discarded, a simple wire of from 100 to 150 feet in height being found to be sufficient.

In 1898 a radical improvement was made in the receiver which immensely increased both its sensitiveness and reliability, and rendered possible the "tuning" of pairs of stations so that interference need not occur. This essential change was brought about by the introduction of a very small transformer between the elevated wire and the coherer circuit.

Of the three advantages which are mentioned above, the reliability was increased by the provision of a direct conducting path to earth, along which any atmospheric electricity could flow gently without affecting the coherer, and

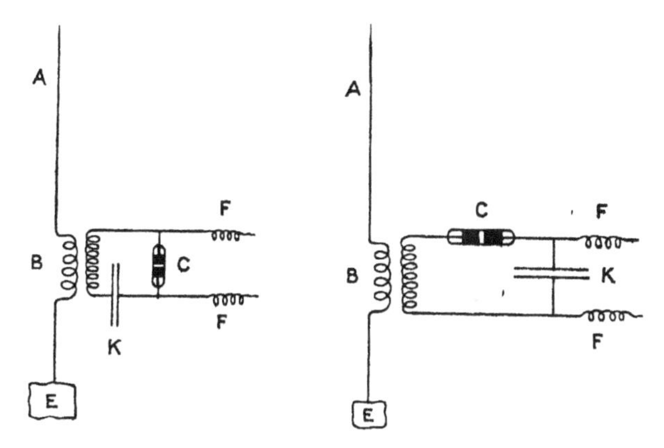

Fig. 25.—Receiving Transformer or Jigger; Two Methods of Connection, to the Coherer Circuit.

A, Aerial Wire; *B*, Jigger; *C*, Coherer; *E*, Earth; *K*, Small Condenser; *F*, *F*, Choking Coils leading to Battery and Relay.

the sensibility was made greater by the increased voltage in the secondary, due partly to its action as a step-up transformer, but mainly to conversion of the coherer circuit into one capable of resonance. This last result of the change was also that which made discrimination between stations possible.

Many and varied types of transformer were tried before one was found on which signals could be successfully received. Ultimately the form adopted consisted of primary coil of about one hundred turns of fine copper wire wound on a glass tube of a few millimetres diameter, with a secondary of about two hundred turns of still finer wire wound in one layer on top of the primary. The primary was connected to the air and earth wires; one end of the secondary was connected directly to one end of the coherer, and the other, through a small condenser, to the other end of the coherer. It was found that, for reasons which will be explained in a later chapter, it was of no advantage to increase the ratio of the secondary to the primary by adding a second or third layer over the first; in fact, that such an addition usually rendered the transformer entirely inoperative. The patents taken out on this subject show that the difficulty was to some extent got over by winding the secondary on a narrow base in layers of diminishing length. The complete solution of the problem was not obtained until it was found that to get the best results there had to be a definite relation between the length of the secondary and that of the aerial wire—that, in fact, the secondary circuit acted as a resonator,

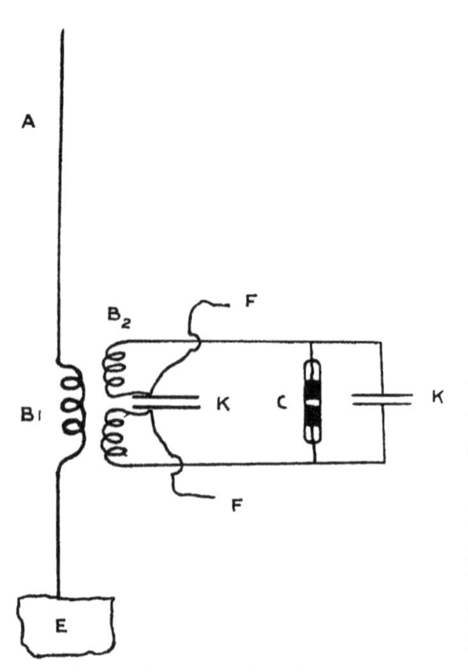

FIG. 26.—JIGGER WITH SPLIT SECONDARY.

A, Aerial Wire; B_1, B_2, Primary and Secondary of Jigger; *C*, Coherer; *K*, *k*, Condensers; *f*, *f*, Leads to Relay Circuit; *E*, Earth.

whose frequency was the same as that of the transmitted waves. We shall return to this subject in Chapter XVI.

A further improvement consisted in splitting the secondary by inserting a small condenser at its middle point; the connections to the relay circuit being made from either side of the condenser, and those to the coherer from the outer ends of the secondary. A still further development was to employ the central part of the secondary as primary, by making connections to two points on the wire separated so as to include only a few turns in the aerial circuit. This method of connection is now called "direct coupling," as opposed to inductive coupling by means of a separate primary and secondary. These transformers are shown in Fig. 27.

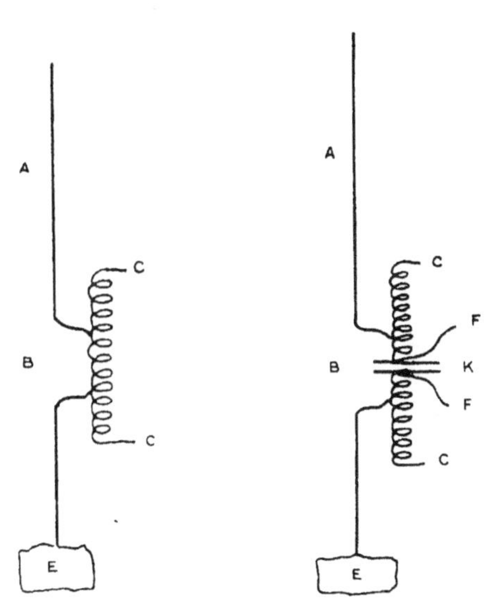

FIG. 27.—DIRECTLY COUPLED AERIAL AND RECEIVING CIRCUIT, WITH AND WITHOUT CONDENSER.

A, Aerial; *B*, Jigger; *C*, *C*, Connections to Coherer Circuit; *E*, Earth; *K*, Condenser; *F*, *F*, Leads to Relay Circuit.

It will be seen that these receiving transformers, or jiggers, as they are frequently called, are of the same type as Tesla's high frequency transformer. The latter is, however, constructed of thick wire, while the former are only successful when made of exceedingly fine wire.

These improvements so greatly increased the sensibility of the receiver that distances of over a hundred miles were possible with a transmitter of the same power as had previously been necessary for communication at ten or fifteen miles.

Lodge.—Sir Oliver Lodge, in 1897, took out a patent for a "syntonic" system of wireless telegraphy, based directly on his own work on the discharge of Leyden jars, and on Hertz's experimental results. The transmitter consisted of two large cones of sheet metal placed with their axes in a vertical line, and having a spark-gap between their apices. In another form of transmitter, a single metal sphere separated by small spark-gaps from the terminals of an induction coil, was used as radiator. Both types produced direct Hertzian radiation, the latter giving waves of very high frequency. The spherical oscillator was partially enclosed in a copper cylinder, open at one end, in order that the rays might be condensed in one direction. The receiver, for use in connection with the large cones, consisted of two similar cones connected through the primary of a small transformer, the secondary of which was connected to the coherer circuit. The dimensions of the transmitter and receiver were adjusted to give

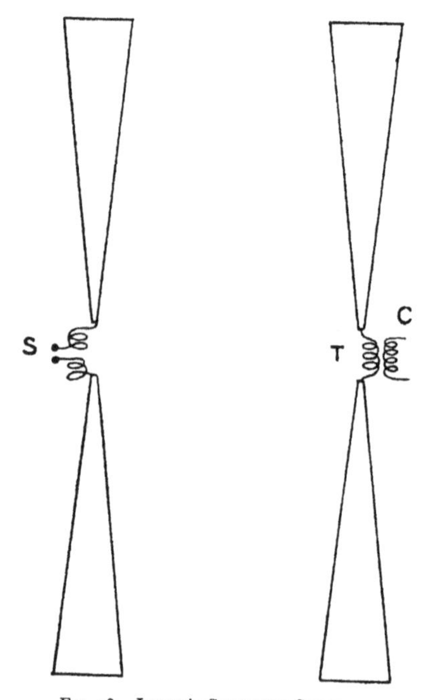

FIG. 28.—LODGE'S SYNTONIC SYSTEM, PRINCIPAL CONDUCTORS.
S, Spark-Gap; *T*, Receiving Transformer; *C*, Leads to Coherer.

equifrequent natural vibrations and therefore resonance. No earth connection was made, as it was desired that the transmission should be purely by means of free radiations. The early conical form of radiator has now given place to the horizontal conductors shown in Fig. 4. Stations in which this latter arrangement has been adopted are now working in various parts of the world.

The systems of which we have just sketched the development, may fairly be said to form the fundamental basis of all more recent work in this branch of electric tele-

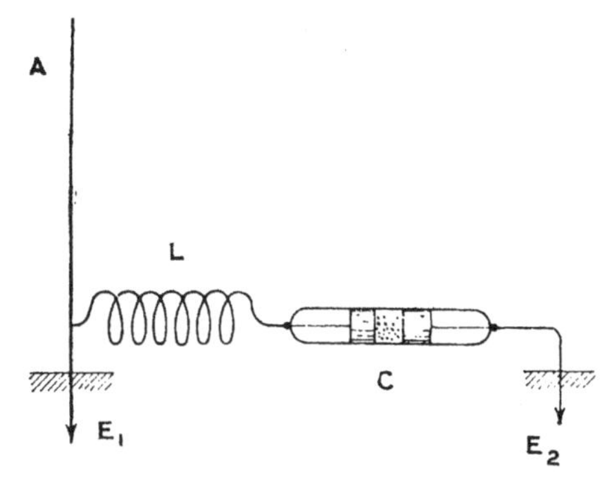

FIG. 29.—SLABY RECEIVING STATION, PRINCIPAL CONDUCTORS, DOUBLE EARTH CONNECTION.

A, Aerial Wire; E_1, E_2, Earth Connections; *L*, Inductance; *C*, Coherer.

graphy. Innumerable inventors have tackled the subject, the result being an ever-increasing crop of patents, of which already some four or five hundred have been granted in this country alone. Many of these are, of course, for trifling changes or for "improvements" long since discarded by the foremost workers in the field, but a large residue consists of devices which are or may be of real value.

The experimental work, for instance, of Professors Braun and Slaby, and of Count Arco, cannot be passed over without notice; particularly as its results are now embodied in the patents of the German Gesellschaft fur Drahtlose

Telgraphie, under the name of the "Telefunken" system. This company works in conjunction with the official services in Germany, and has probably a larger number of stations than any other in existence. It was reported at the annual meeting in 1905, that 518 installations had been erected.

The Slaby-Arco was apparently the first system in which an aerial wire connected to both terminals of the induction coil was used. The wire was arranged in the shape of an arch with a spark-gap at the lower end of one side. If the two sides of the arch be of equal electrical dimensions, no radiation takes place; but if the capacity or inductance of one be considerably greater than that of the other, the circuit is unsymmetrical and becomes a good radiator. It is an advantage to connect one of the terminals to earth. A somewhat similar acrial is cmployed by Maskelyne. This type of aerial has some advantages over a single wire, but is at the same time more troublesome to erect. It appears to be no longer in general use. Another method employed by Slaby was to use one aerial wire, but two earth connections; these were connected by a wire through the receiver. The receiver is in this case tuned by the insertion of an inductance in series with it, and only responds to waves of a certain frequency. Atmospheric disturbances

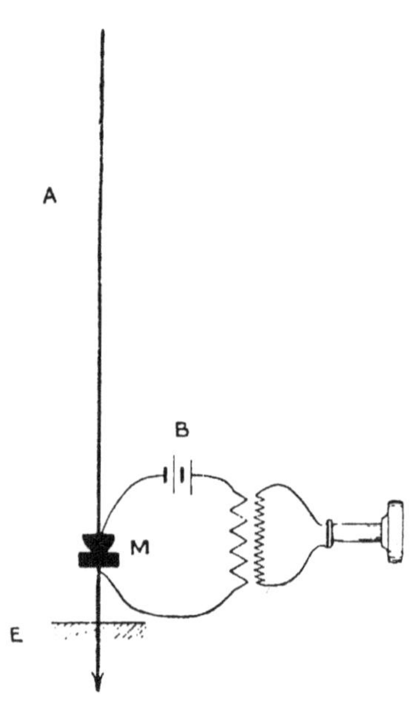

FIG. 30.—MICROPHONE AS DETECTOR IN CONJUNCTION WITH TELEPHONE.
M, Microphone; *B*, Battery; *A*, Aerial Wire; *E*, Earth.

do not disturb it unless they are exceedingly violent, nor do transmitters of a different frequency.*

For transmission, the spark-gap between the terminals of the induction coil is at the middle of the horizontal wire, one end of which is earthed through a condenser and inductance, while the other is connected to earth through an inductance only. In cases where a record of the message is not required, a combination of microphone and telephone forms a very sensitive receiver. The microphone acts, as Hughes showed, as a coherer, and the click produced in the telephone renders the signal audible. A telephone in conjunction with some type of high frequency current detector is used in many modern systems—among others by Marconi, with his magnetic detector; by Fessenden, with an electrolytic responder; and by De Forest and many more in connection with various species of coherers or microphones.

---

* Slaby, "Fortschritte der Funkentelegraphie," *Zeitschr. des Vereines Deutscher Ingenieure*, July 27, 1901.

CHAPTER III.

APPARATUS USED IN THE PRODUCTION OF HIGH FREQUENCY CURRENTS.

THE methods of producing currents suitable for the transmission of electrical energy over the earth's surface without the use of a return wire, *i.e.*, without any wire at all between the stations, have been sketched in Chapter II. It is now necessary to describe in some detail the types of apparatus which are employed for this purpose. It is clear that as there is no return wire the only practicable way of distributing the energy necessary to telegraphy is by means of alternating currents. Since the energy transmitted depends both on the current and voltage, it is necessary, in order to avoid the use of excessively large capacities, to obtain large currents while using small quantities of electricity by increasing the rapidity of the motion of the electricity. High frequency of alternation and high voltage are therefore employed almost universally in wireless telegraphy.

In an alternating current generator of the usual power station type the number of alternations per second is governed by the number of pole pieces and the rate of revolution. For electrical and mechanical reasons it seems impossible to increase either of these quantities indefinitely, as the losses from eddy currents and hysteresis become very great, and there is also danger of the machine flying to pieces from the high speeds necessary. Mr Tesla in

1893 obtained a current of several amperes at frequencies of from ten to thirty thousand periods per second; ten years later Mr Duddell constructed a generator which gave about a tenth of an ampere, the frequency being 120,000 periods per second. Great as these frequencies are, they are still hardly high enough for working effectively with the comparatively small capacities which are convenient in wireless telegraphy; frequencies ten or a hundred times as great are more suitable in almost every case.

The accompanying illustration (Fig. 31) shows a gene-

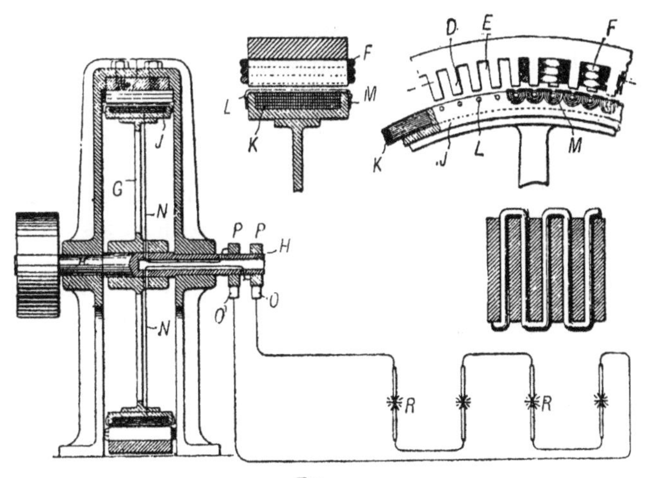

FIG. 31.

rator constructed by Mr Tesla in vertical cross-section, with some details. As will be seen, the machine comprises an annular magnetic frame, the interior of which is provided with a large number of pole pieces D. Owing to the very large number and small size of the poles and the spaces between them, the field coils are applied by winding an insulated conductor F zigzag through the grooves, as shown, carrying the wire around the annulus to form as many layers as is desired. In this way the pole pieces D will be energised with alternately opposite polarity around the entire ring. For the armature, Mr Tesla employs a spider carrying a ring J, turned down, except at its edges, to form a trough-like receptacle for a mass of fine annealed iron wires K, which are wound in the groove to form the core proper for the armature coils. Pins L are set in the sides of the ring J and the coils M are wound over the periphery of the armature-structure and around the pins. The coils M are connected together in series, and these terminals N carried through the hollow shaft H to contact-rings P P, from which the currents are taken off by brushes O. In this way a machine with a very large number of poles may be constructed. It is easy, for instance, to obtain in this manner three hundred and seventy-five to four hundred poles in a machine that may be safely driven at a speed of fifteen hundred or sixteen hundred revolutions per minute, which will produce ten thousand or eleven thousand alternations of current per second. Arc lamps R R are shown in the diagram as connected up in series with the machine.

To produce such currents, the fact that if a condenser be discharged through a circuit of low resistance the current is not continuous but oscillates to and fro a number of times before dying out, has also been made use of. To obtain the high voltage necessary to charge the condenser, two methods are available. The first is to use an electrostatic machine; but on account of the small output of even a large machine, this method has not been used in practical telegraphy. The second method consists in transforming a current of low voltage into one of very high voltage by means of some type of induction coil or transformer. In the induction coil an intermittent unidirectional current in the primary coil induces an alternating current of high voltage in the secondary. On making the primary circuit the current in it increases comparatively slowly, inducing a reverse current in the secondary of low voltage, since this depends directly on the rate of change of current in the primary. On breaking the circuit the primary current ceases almost instantaneously, thus inducing a momentary current of high voltage in the secondary. It is this current which in the simpler systems of wireless telegraphy is used to charge the aerial wire up to the voltage at which the electricity sparks across the gap to the earth wire. A frequency of one or two thousand interruptions per second is the maximum which has been as yet attained, hence the current in the aerial consists of short trains of oscillations of very high frequency following one another at

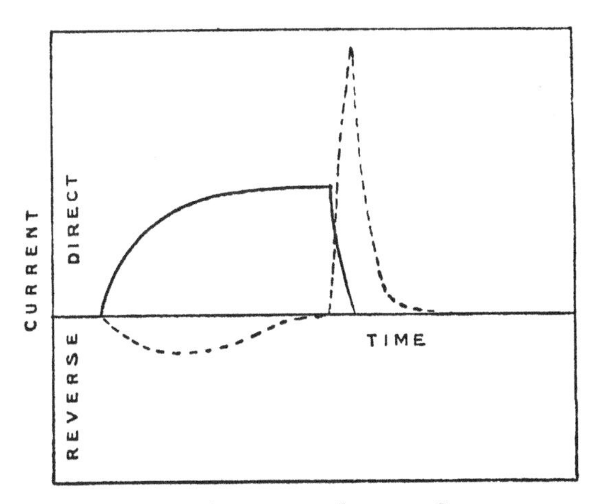

FIG. 32.—ACTION OF AN INDUCTION COIL.
Horizontal Distance represents Time. Vertical Distance, Current. The full line represents Current in Primary, and the broken line the Current induced in Secondary.

comparatively long intervals, *i.e.*, each train of waves may occupy $\frac{1}{100000}$ of a second, while the interval before the next commences may be $\frac{1}{1000}$ of a second, thus the blanks may be a hundred times as long as the active periods.

Many improvements have been made during the past few years in the construction of the induction coil, the most important being due to the recognition of the fact, which Tesla pointed out in 1893, that the best results can only be obtained if the natural periods of oscillation of the primary and secondary circuits have some simple relation to one another. To attain this end it may be constructed so that the electrical dimensions of the two circuits should be such as to give the same frequency of vibration. This can be done by properly adjusting the capacity and inductance of the circuits so that the square roots of the products of these quantities for each circuit are equal. One of the most successful pieces of apparatus for this purpose is the Grisson interrupter and resonance coil. Electrolytic condensers are used in order to obtain large capacity with moderate sized apparatus, and a revolving commutator controls the circuits so that the break takes place at a moment of zero voltage, while the circuit is *made* when the conditions are such that an instantaneous rush of current takes place through the primary. This interrupter therefore appears to act on the rapid rise of the current instead of on its fall as is usual in other apparatus for a similar purpose. Electrolytic interrupters of the Wehnelt type have come into use recently, their nominal advantage being that their rate of sparking is about ten times as great as that of the hammer break; they are also more easily kept in order than any form of break in which mercury is used as an electrode. The increased spark frequency may be of use in Rontgen ray work, but it is quite unnecessary and may even be harmful in wireless telegraphy. The Wehnelt interrupter consists of a glass cell containing weak acid, in which are placed two electrodes. One of these, the anode, is a small platinum point, the other is a large lead plate. Its action is discussed in Chapter VII.

It is fallacious to suppose that a very high rate of sparking, such as that given by the Wehnelt break, is necessary for rapid telegraphic work. For speeds much higher than those generally attained in wireless telegraphy, interruptions at the rate of about 100 per second, as given by the hammer break, are quite sufficient. Thirty words a minute, which is considered rapid on any system of telegraphy when the sending is done by hand, means the letter V repeated 150 times in one minute. Now V, on the Morse code, is represented by three dots and one dash. In order to transmit this letter correctly, only about ten sparks, or rather vibrations of the hammer, are required; one for each dot, three or four for the dash, four more which might have occurred during the blank spaces had the letter been a different one. Now 150 Vs per minute is 2.5 per second, hence we require 25 sparks per second. So with 100 sparks per second, as may be given by the hammer break, we have four times as many as are necessary; a factor of safety of four. As a matter of fact, the author has received correctly Vs sent at the rate of 60 words—*i.e.*,

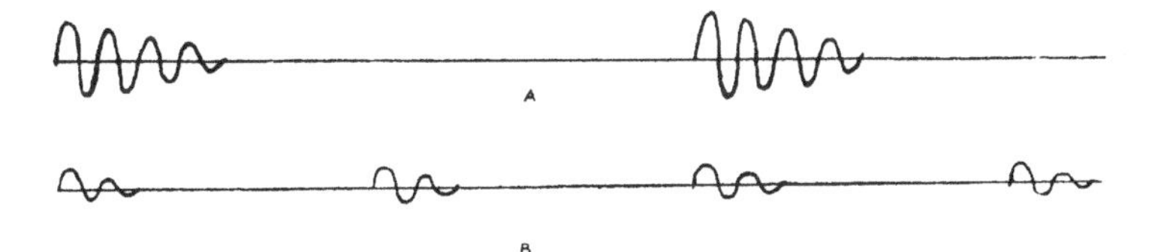

FIG. 33.—TYPES OF WAVES RADIATED.

A, Wave trains of considerable amplitude; *B*, Wave trains of less amplitude occurring more frequently. *A* is generally more useful than *B*. (*N.B.*—The actual intervals between wave trains are much longer in comparison to the length of the train than those shown.)

300 Vs—per minute by an automatic key in series with an ordinary hammer break. It is essential to good transmission that every spark should be a good one, but the number is of less consequence. The reason for this is, that almost all detectors depend for their action on the maximum voltage or current received, and not on the integral current as read on an instrument; and the increase of frequency of sparking does not affect either the maximum voltage or maximum current, but merely increases the integral current. Increase of sparking rate, beyond a certain value, must be definitely disadvantageous, since the time required for the core of the coil or transformer to go through its cycle of magnetic changes, is by no means

infinitesimal; and at high frequencies the hysteresis and eddy current losses become very great.

One advantage of the hammer break, which certainly counterbalances some of its disadvantages, is the ease with which two important adjustments may be made. These are (1) the rate of vibration, and, which is more important, (2) the duration of the contact. By the mere turn of a screw, the length of time during which the current is flowing in the primary, and therefore the maximum magnetic induction in the core can be regulated to suit the load. The chief disadvantage is the rapidity with which the contacts wear when heavy currents are used, and the fact that it is impossible to use more than about 25 volts without producing arcing.

Mercury breaks are of two types; in the older form mercury in a small dish is connected to one electrode, while the other is attached to a plunger which dips into the mercury and is periodically withdrawn by a reciprocating mechanism. This form is practically obsolete. In more modern breaks, one electrode is a jet of mercury driven by a turbine pump against a species of revolving toothed wheel which is connected to the other electrode. The teeth are tapered, so that by adjusting the position of the jet which plays against them, longer or shorter contacts may be made. The chief disadvantage of this type is, that the mercury becomes churned up with the petroleum or other insulating liquid in the jar, and rapidly forms a sticky mess which must be periodically cleaned out and purified. In a laboratory, with conveniences for chemical work, this is not a serious drawback, but in a wireless telegraph station—possibly one room in a cottage, many miles from even a village—it is a great objection, especially as while the apparatus is being cleaned, there is no possibility of answering a call, or of sending an urgent message. For these reasons, mercury breaks have not entirely superseded the other forms of interrupter mentioned, though they have frequently been used on account of their uniformity of action for short periods; *i.e.*, before the mercury and oil have become appreciably mixed.

In wireless telegraph stations, where continuous current is used, the interrupters in most general use are the hammer and the mercury turbine, but interrupters are being dispensed with entirely in many stations by the adoption of alternate current machinery. In the earlier stations the source of electric energy was usually a battery of, say, fifty large dry cells, used perhaps in combination with a portable accumulator. Nowadays it is more common to have a small oil or petrol engine and alternator in conjunction with some high tension transformers, instead of a battery and induction coil. The frequency of the alternator is not related to the frequency of oscillation, and only indirectly to the number of sparks per second. The actual rate of sparking is not in general exactly twice the frequency of the alternator, as might have been expected from the occurrence of a positive and a negative maximum each period, but may be much greater; in fact, there may be as many as ten sparks per alternation. This is probably due to electric surging in the circuit; *i.e.*, oscillations of shorter period superposed on the wave given by the alternator. The phenomenon depends on the induction in the leads from the transformer to the spark-gap and the capacity charged. Inductance must be introduced in these leads, otherwise an arc would be formed which would be useless for the creation of jigs. Inductance, however, checks the supply of current when the spark has commenced and the dropping of the voltage causes its sudden but oscillatory extinction. The phenomenon may thus occur several times during each period of the alternator. This subject has been investigated by Professor Fleming, by means of an ingenious spark counter which he has devised for the purpose.*

In stations run by alternating machinery the current is generated at a low voltage, and is transformed to the very

* "Electric Wave Telegraphy," p. 157, J. A. Fleming (Longmans).

high voltage necessary by means of one or more transformers. These must have high ratios of transformation, and of course very good insulation. A potential of about 30,000 volts or more is common in such work, though some inventors, notably Professor Fessenden, are understood to be using lower voltages.

CHAPTER IV.

DETECTION OF SHORT-LIVED CURRENTS OF HIGH FREQUENCY BY MEANS OF IMPERFECT ELECTRICAL CONTACTS.

IN order to detect the existence of oscillatory currents in the circle of wire which he used as a resonator, Hertz made a small break in the continuity of the conductor and observed the sparking which took place there. This method is obviously not a very sensitive one, as a considerable voltage must exist across the gap before a spark will pass, even if the gap be very small.* No practical system of telegraphy could therefore be based on the use of so insensitive a detector.

Ordinary telegraphic instruments are totally inapplicable to the detection of high frequency currents, since even the most sensitive requires that a unidirectional current should pass through it for about $\frac{1}{500}$ of a second before it will act. The currents employed in wireless telegraphy are, however, oscillatory, and last perhaps, $\frac{1}{100000}$ of a second; to detect them it is thus necessary to use an instrument which will be actuated by a current lasting only the 200th part of the time required by an instrument suited to wire telegraphy. And again, the fact that the currents are oscillatory prevents the employment of the electro-magnets which are so general in apparatus for wire telegraphy. Even the telephone is unsuitable because of its inductance and also

* See Tables, Chapter XXI.

on account of the inability of the ear to detect tones of very high frequency.

For wireless telegraphy therefore, not only was a new type of transmitter requisite, but also an entirely new method of detecting the transmitted currents and so receiving the signals.

The first practical detector of oscillatory currents of high frequency was the loose contact between conductors named by Hughes the microphone, and shown by him to have the power of detecting both mechanical and electrical vibrations. If the loose contact be between two pieces of carbon, the change in resistance is only momentary, the effect is therefore not detectable on a galvanometer, but is easily observed by means of a telephone since the latter is so much more rapid in action. With metallic contacts a permanent change of resistance takes place which may be indicated by a galvometer or any type of telegraph instrument in circuit with the loose contact and a local battery. The contact does not, however, restore itself to it soriginal state of sensibility automatically but remains cohered, or welded, until the pieces of metal are separated by mechanical force. The actual force which must be applied in order to separate two pieces of metal which have been electrically cohered has been measured by Dr P. E. Shaw.*

Coherers consisting of only two conductors in loose contact have the disadvantage of being very easily disturbed by mechanical vibration, and are thus unreliable and of little use for telegraphic purposes. The carbon microphone has, however, been used by Braun and others with considerable success as a detector.

The sensitiveness of the loose contact to the action of electric currents was rediscovered by Calzecchi-Onesti, Lodge, and Branly. The last named employed the form which has been developed into the coherer of practical telegraphy. Branly's tube had an electrode at each end, and

* *Phil. Mag.* March 1901 and August 1904.

was filled with a quantity of fine metal filings. The uncertainty of a single point contact was thus obviated and the disturbing effects of mechanical vibrations were minimised. In its original form its action was somewhat erratic at times ; Marconi has, however, developed from it an instrument which, when properly treated, is as reliable in its action as any other telegraphic apparatus. Early experimenters spoke of the coherer as delicate and unreliable—a criticism which was in reality a confession of ignorance of its proper construction and conditions of use, and though for some purposes it has been superseded by magnetic and other detectors, it is still in use at a great number of stations of various companies. Mr Marconi reduced the size of the

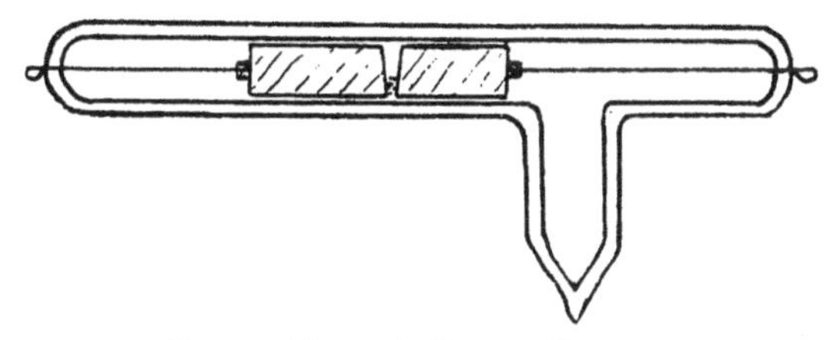

FIG. 34.—MARCONI'S FORM OF COHERER.

tube until the space containing the filings was merely a small disc-shaped cavity less than a millimetre long and about five millimetres in diameter between two silver plugs. The filings filled only about one-third of the cavity and consisted chiefly of nickel, a trace of silver being added to increase the sensibility, and the tube contains a partial vacuum.

At the author's suggestion the filings were sifted through pieces of fine silk gauze to obtain uniformity in size. This construction was of course only arrived at after many experiments with tubes of other dimensions and containing filings of many other metals. The advantage of the coherer for wireless telegraphy is that it remains cohered until the filings are shaken up again. The momentary oscillatory current thus records itself in a permanent reduction of the resistance of the coherer, which continues so long as the tube is not shaken. The introduction of the coherer in the path of the oscillatory currents therefore provides a means by which any ordinary telegraphic instrument can be actuated, and will respond only on the arrival of oscillations at the coherer. The coherence lasts while the current from the local battery flows through the circuit and causes the telegraph instrument to indicate its presence. If the effect on the coherer had been merely transient instead of semipermanent, it would not have been possible to use a recording telegraph instrument, or indeed any kind of apparatus whose action is controlled by electro-magnets.

Since the coherer does not return to its original sensitive state automatically, it is necessary to arrange that it should be shaken immediately after a signal has been recorded. This of course involves the use of a mechanism which acts in conjunction with the recording instrument and administers a tap to the tube sufficient to decohere the filings. The complete receiver is thus somewhat more complicated than if a self-restoring detector were used, but it has the advantage that it is capable of producing a permanent record, on a Morse tape, of the signals received. Where a self-restoring detector is used the signals must as a rule be read by the sounds they produce in a telephone receiver.

Though many variations have been proposed by various experimenters, the utility of the type of filings coherer designed by Mr Marconi, is proved by its continued use at the stations of many companies in Europe and America. The mode of action of the filings coherer has been the subject of much discussion. Many of the earlier experimenters appear to have fallen into the curious logical fallacy of supposing that because electric waves caused coherence they were therefore essential. The fact is that the only necessary condition is that a certain voltage be applied to the tube of filings; whether it be oscillatory or continuous is probably a matter of indifference. If a slowly but continuously increasing voltage be applied to the terminals of a coherer, its resistance will diminish just as rapidly, when a certain point has been reached, as if electric waves had arrived. After all, electric waves are merely rapid changes of potential, and it is not the rapidity which causes coherence but the difference of potential. At my suggestion Dr W. H. Eccles made a very thorough experimental investigation of the subject, and found (as I had always supposed) that oscillations have nothing to do with the matter, *i.e.*, that they are by no means essential to coherence. His results are so important both from the theoretical and practical points of view, that I shall later on quote somewhat fully from his paper.

In Faraday's well-known method of showing the lines of force of a magnet, small particles of iron form chains when placed in a magnetic field. If they are lying on a comparatively rough surface, such as paper, gentle tapping, which lifts them momentarily into the air and thus allows them to set themselves freely along the line of force, facilitates the action. Robertson has shown that small conducting particles, such as sawdust, behave in an exactly similar manner under the influence of electrostatic force. Their ends become charged, and as the ends of neighbouring particles, in the direction of the force, are oppositely charged, they attract one another and form conducting chains. The final result is a greater conductivity in the whole mass. It is important to note, from a practical point of view, that intermittent mechanical shocks facilitate the formation of chains, just as in Faraday's experiment. This explains the well-known influence of the adjustment of the tapper in a wireless telegraph receiver.

I have noticed that if the voltage in the local circuit of a coherer be increased to a certain point by means of a potentiometer arrangement, the sensitiveness of the apparatus is much increased. If the steady potential applied be nearly large enough to cause coherence, the slightest rise, due to the waves, is sufficient to determine the result.

With this preliminary I shall now quote from Dr Eccles' paper.*

"*Preliminary.*

"The investigation described in the following pages† was commenced early in 1900 with the object of revising what was then known of the relative influences of oscillatory and of steady P.D.s in promoting coherence in filings coherers. It was hoped that evidence might be obtained which would decide definitely whether the action of electrical vibrations was in character different from or similar to that of steady E.M.F.s. The importance—from several points of view—of this discrimination is manifest; and that the difference, if any, has not, up to the present, been pointed out, nor, if non-existent, been proved so, is shown by an examination of the published records of the research on the subject.

"Many observers have assumed or have suggested that the events leading to coherence among a mass of particles were identical in the two cases mentioned. They have often expressed this assumption implicitly by postulating electrostatic attractions between the particles. To go a long way back—in 1870—Varley, in describing to the British Association Meeting his lightning protector for telegraphic apparatus, suggested that the bridging over of the gap resulted from the electrostatic attractions produced by large E.M.F.s among the particles of the conducting powder placed between the terminals. Even earlier, in 1850, Guitard made similar remarks regarding his experiments on dusty air. And since then, other experimenters working on our or on kindred subjects have given similar vaguely expressed views. Again,

---

* *Electrician*, August 23 and 30, 1901, by kind permission of the publishers.

† Many of the tables accompanying this article were obtained from coherers forming part of the stock of the Marconi Company. It is only by the company's kind permission that the writer is now enabled to publish them.—W. H. E.

Branly, in his classical paper,* having observed that the application of the E.M.F. of a battery of mercury sulphate cells produced conductivity in a mass of filings, and that the effects so produced were identical with those caused by the incidence of electric radiation on the filings, therefore asserted that the effect in both cases was one of E.M.F. merely. This view of the matter seems to have been accepted tacitly by many subsequent workers. But in no case has special care been taken by experimenters to avoid the creation in the coherer circuit of electrical disturbances of an oscillatory character. So that it is at present, to the author's knowledge, the opinion of several eminent authorities that the electrical surgings set up in the coherer circuit on the sudden establishment of a connection with a source of E.M.F.—as by any ordinary contact device used in conjunction with a battery—play an important part in the production of coherence. The exact function fulfilled by such electrical surgings has by no one been exactly outlined, but it is probably imagined by most to consist in a modifying of the mechanical arrangement of the particles. Such an action was broadly indicated by Lodge in the general case of a filings coherer submitted to electro-magnetic radiation when he spoke in 1899† of the washing of the ether waves round the particles aiding their orientation.

"With regard merely to this question of movements among the particles, considerable divergence of opinion has, apparently, always existed. Of course, the effect of tremors in assisting coherence and of shocks in destroying it has been noticed by all observers, and has, by the majority, been ascribed to actual relative motions among the particles. Calzecchi, in 1884, used a mass of filings heaped between two brass plates in series with a cell and galvanometer as a tremor detector, and noticed‡ that even sounds affected the conductivity of the mass. He calls the arrangement, indeed, a 'microseismic indicator.' On the other hand, Branly,§ led by his experiments on compressed masses of filings and on filings embedded in blocks of resin, in introducing his theory, postulates that motion of the particles

* *La Lumière Electrique*, May and June 1891.
† Royal Institution Lecture, February 24, 1899.
‡ *Nuovo Cimento*, September 1897.
§ *La Lumière Electrique*, June 1891.

is impossible. Pasquini * recommends the greatest possible freedom of motion for the particles. Arons † finds that Canada balsam, which must of necessity greatly hinder the motion of the particles, did not interfere with the efficiency of the coherer's action. Several observers have investigated the effect on conductivity due to the agitation caused by sound waves. But Branly,‡ in 1898, still recommends compression in using filings coherers. Tommasina,§ however, in 1899, and Sundorph ‖ and Malagoli ¶ in the same year, demonstrated the formation of lines or chains of filings; and these lay stress on the occurrence of motion among the particles.

"During the past year (1901) little has been added to our knowledge of the action of unidirectional E.M.F.s on filings coherers. In April, however, while the author was engaged on his experiments, Blondel and Dobkevitch communicated ** the results of their investigations. They had found that different coherers submitted in turn to gradually increased E.M.F.s cohered respectively at voltages which approximately indicated their sensitiveness. They called that P.D., which, steadily applied, produced coherence, the 'critical voltage.' They pointed out that for the purposes of wireless telegraphy the E.M.F. of the battery in the coherer-relay circuit (plus possible E.M.F.s due to the inductance of the circuit) must not exceed this critical value, and that the E.M.F. added by impinging waves should enable the critical value to be passed. The present writer, however, shows in this article that the coherers with which he worked—and they were of considerable number and variety—gave 'critical voltages' which were very indefinite, these being dependent, as they seemed, on the tremors to which the coherer, being examined, was liable, even under the best working conditions. Moreover, it will also be here shown that there is no true 'critical voltage,' the degree of coherence produced by even the smallest

---

* *Nuovo Cimento*, March 1898.
† *Wied. Ann.* No. 7, 1898.
‡ *Comptes Rendus*, December 26, 1898.
§ *Ibid.* May 15, 1899.
‖ *Wied. Ann.* No. 7, 1899.
¶ *Nuovo Cimento*, October 1899.
** *Comptes Rendus*, April 23, 1900.

practicable E.M.F.s being perfectly measurable with an instrument of suitable sensitiveness. It will be seen that exact information on the various questions which arise in connection with the filings coherer is scanty in the extreme. Much more attention has been paid to the single-contact coherer. But as this article deals only with coherers made with filings, it is not necessary to revise the work which has been carried forward on the more carefully studied part of the subject.

"The phenomenon of coherence, whatever it may consist in—whether it is, as Lodge affirms, a welding together of the two surfaces in pseudo-contact, a reaching of the forces of cohesion across the separating space; or whether it is, as Branly considers, a modification of the dielectric occupying the minute space between the conductors concerned; or, as Bose has proposed, an effect depending on allotropic change in the material of the conducting particles; or, as others believe, one depending on the existence of surface films of condensed gases or moisture—this phenomenon of pure coherence is, in all probability, greatly disguised and complicated in a filings coherer by the preliminary and purely mechanical actions of the particles. That the phenomenon of coherence proper, then, cannot be advantageously studied amid these probable complications, need scarcely be pointed out. This view of the case seems also to the writer to account in a great measure for the extraordinary conflict of opinion which has obtained, since Branly's discovery, with regard to various qualities, such as the relative sensitiveness, of coherers made with filings of different metals. To cite only a few cases, it is notable that Marconi, in his patent specification of March 1897, states that he uses a proportion of silver filings to increase the sensitiveness of his nickel coherers, and that the continued addition of silver greatly increases the sensitiveness; while Lodge,* in the same year, remarked that the noble metals are not suited for coherers. Dorn,† in 1898, confirmed Lodge, and with the noble metals placed nickel, adding that only oxidisable metals are sensitive. Branly‡

* *Electrician*, vol. xl. p. 87.
† *Wied. Ann.* No. 9, 1898.
‡ *Comptes Rendus*, December 26, 1898.

then announced that, on the contrary, the noble metals are extremely sensitive; and Blondel and Dobkevitch, in their paper quoted above, seem to agree with Branly when they state that the less oxidisable metals give lower 'critical voltages.'

"Such confusion has arisen, the writer believes, out of the great complexity of the numerous conditions which influence the mechanical and electrical behaviour of the filings in a coherer. These conditions may be seen to be dependent, broadly, on the state of rest or of motion of the particles and on their relative positions at the moment of coherence; on the nature, probably, of the material of which the filings consist; and on the size and shape of the filings. Evidently the size and shape of the particles each depend on the manner of production—that is, on the kind of file used and the mode of using it, and must also be, for different substances, to some extent dependent on the elasticities, the tenacity, and other properties of those substances. The mechanical circumstances alone are thus extremely involved. But to co-ordinate the experimental data accumulated by the author it was early found necessary to construct a working theory. This theory—which is sketched later—springs naturally from the results of the first few series of experiments. That these experiments were necessary for the framing of any hypothesis intended to correlate the movements of the particles with the E.M.F.s applied is evident. For, if oscillatory discharges were proved necessary for efficient coherence—that is, in the light of Tommasina's experiment for the formation of numerous chains of cohered particles—then the effect on the movements of the particles of displacement currents in the dielectric must be taken into account; while if, on the contrary, steady E.M.F.s were shown sufficient, then other electro-mechanical connections must be proposed.

"*Experimental.*

"To confirm or correct and to amplify the very meagre information available on the effects of steady E.M.F.s on filings coherers the following experiments were projected during the spring of 1900. They were made possible by the kindness of Dr J. Erskine-Murray, F.R.S.E., in whose laboratory they were commenced, and whose advice and assistance the writer had the good fortune to have always at call during the first phase of the work. The plan proposed was to submit to the influence of a steadily increased P.D. each of a large number of coherers of known degrees of sensitiveness, and to observe at what E.M.F. (if any) coherence was produced. Through Dr Murray's kindness, the writer had access to a very large stock of coherers which had been tested by the use of signals received on an air-wire, and whose general character and whose sensitiveness were fairly well known.

"These coherers were submitted to E.M.F.s supplied by a potentiometer arrangement. To avoid the possibility of abruptness in the changes of potential, a copper-sulphate solution potentiometer was adopted. This consisted merely of a long graduated glass tube containing dilute copper-sulphate solution, with one movable and two fixed electrodes. The fixed electrodes were of copper, and introduced and led away respectively the main current supplied by a battery of 10 volts. The movable electrode was also of copper, but was exposed to the electrolyte only at its extremity, the shank by which it was moved being insulated. Its position was read on the graduations of the tube, the value of the graduations in volts being roughly found by a separate experiment. One of the fixed electrodes (earthed as a rule) and the movable electrode formed with the intervening electrolyte a portion of a circuit which also included a moving coil galvanometer, a resistance of 12,000 ohms, and the coherer which was to be examined. By this means could be applied to the coherer a gradually increased E.M.F., ranging from zero to 10 volts, and that in so gradual a manner that there arose no probability of the creation of electrical surgings in the circuit external to the coherer. It might be added that a very few preliminary experiments showed the advisability of including in the coherer circuit a high resistance: its inclusion prevented the passage of any current considerable enough to damage the coherer. The coherers used were all of one type, were made by the same man from the same materials, and consisted of nickel filings between silver electrodes about 0.3 mm. apart, sealed in a vacuous glass tube.

"In every case coherence was found to take place with

suddenness. Immediately coherence was indicated by the violent throw of the galvanometer mirror, the main current of the potentiometer was switched off and the position of the movable electrode read on the glass tube and recorded. Then the coherer was restored to its normal condition of high resistance by gentle tapping continued for a few seconds and the experiment repeated. It was found that the E.M.F. required to cohere the same tube varied as much as from 20 to 60 per cent. on successive occasions. The movement of any occupant of the room below affected the value of the figure obtained appreciably, while gentle agitation of the tube usually ensured coherence at quite low voltages. The effect of rougher shocks was, as Branly and others had shown previously, very irregular, and, as observation of the galvanometer mirror indicated, sometimes took the form of a rapid succession of coherences and decoherences. As regards the behaviour of different coherers all exposed to identical circumstances, it was found, even when the greatest efforts were made to protect the tube from mechanical disturbances, that no law could be proposed to connect the E.M.F. at which coherence occurred with the sensitiveness as ascertained by testing on an air-wire—excepting in those two extreme cases in which the coherers were either extremely sensitive or extremely inert as regards signalling. In these two cases, indeed, low P.D.s for the one (from 0.1 to 0.5 volt) and high P.D.s for the other (from 6 to 10 or more volts) were the rule. In the case of very sensitive tubes the movements of the galvanometer mirror were extremely erratic at about the cohering voltage. No external cause could be definitely assigned to account for these vagaries, but they perhaps arose from mechanical disturbances originated at a distance and transmitted through the walls of the building to the apparatus table. A selection of the figures obtained is given in Table I. Each figure is the mean of three distinct observations obtained under circumstances as similar as possible. With these figures may be compared the results of independent tests made for commercial purposes on each coherer by examining its behaviour in receiving signals transmitted from a distant station.

"TABLE I.

Character of Coherer by Radiation Test.	Cohering Voltage.	
	Liquid Potentiometer.	Wire Potentiometer.
A. Too sensitive	0.05 0.2	0.15 0.1
B. Very sensitive	0.6	0.8
C. Moderate	2.0 0.4 3.5	1.2 1.2 4.1
D. Hard	1.1 5.2	0.7 3.6
E. Very hard	0.3 4.2	0.5 5.0
F. Too hard	10+ 5	10+ 8

"The liquid potentiometer was now discarded and a simple form of wire potentiometer installed in its place. The experiments outlined above were repeated on the same coherers, continuous contact between the slider and the potentiometer wire being purposely avoided. It will be seen from Table I. that figures were obtained which, allowing for the known aberrations, were for each coherer similar to those obtained with the electrolytic potentiometer.

"So far, on the whole, mere confirmation of the results of other observers had been obtained. But now, it may be admitted, all doubts as to the possibility of coherence by the application of absolutely non-oscillatory E.M.F.s had, by the use of the liquid potentiometer, been removed. Moreover, it was now fairly well established that any electrical surgings produced by sudden connection with the source of E.M.F. had little or no influence in compelling coherence beyond that which their maximum voltage gives them. In other words, it appeared to be sufficiently demonstrated that coherence was an effect caused merely by P.D. It should be observed, however, that it is always possible, even when the applied P.D. increases slowly and continuously, that electrical surgings may be created by the first small discharge which takes place through the coherer along that line of particles which happens at the moment to offer the least resistance, and that the initial slight surgings so produced facilitate similar action along other lines of particles, till by a sort of cumulative process a large proportion of the whole mass of filings becomes involved in the process of coherence.

"Unfortunately, the vagarious numerical results so far obtained indicated so little by their magnitude the characters of the coherers from which they were derived, that some statistical mode of attack seemed necessary to detect the relation which it must be presumed exists between sensitiveness and cohering voltage. The most obvious method was to obtain the cohering voltage of each coherer in a large number of successive trials and take the mean. This process was adopted to this extent: that thirty values were obtained from each of six coherers of medium sensitiveness. The circumstances of each test were made as identical as possible, and the decoherence preceding each test consisted of half a minute's tapping by a bell-trembler. The results of striking an average showed that there was a decided tendency for the figures to cling round a particular mean for each coherer, the values of these means lying all between 1 and 4 volts, but not agreeing strictly in their order with the known order of sensitiveness of the coherers used. No coherers that were extremely sensitive or extremely non-sensitive were submitted to this tedious method, for a single trial always picked out such tubes sufficiently definitely.

"The laboriousness of this procedure prohibited its application to any large number of coherers. It therefore seemed necessary to vary the method of attack. Doubtful of finding any practicable mode of securing perfect freedom from mechanical disturbance, the writer next endeavoured to secure that the degree of mechanical disturbance should be uniform. This was more or less effectively managed by fixing the coherer to a framework kept in gentle agitation by an attached rapidly-vibrating bell-trembler. Unfortunately, another long series of measurements conducted on the same lines as before showed that little had been gained; the figures obtained from the same coherers, though being all smaller than by the previous method, were as irregular in value, and, apparently, as little related to the behaviour of the coherers under receipt of radiation as those set out in Table I.

"It now occurred to the writer to reduce the time of a

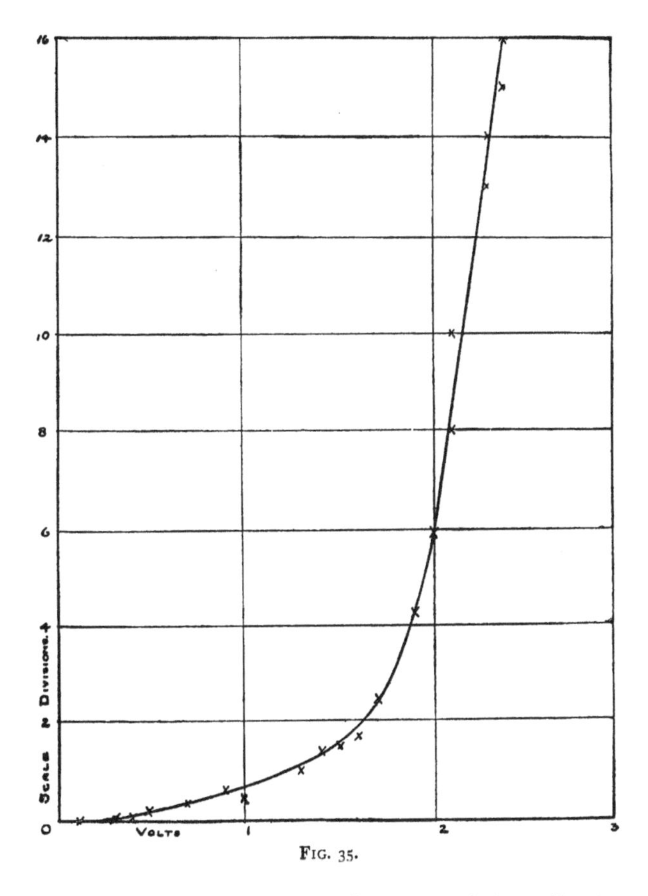

FIG. 35.

single trial to the fraction of a second by allowing the hammer of a trembler to strike uninterruptedly the coherers under examination. During each quiescent stage between the blows coherence might take place and a current pass through the galvanometer. Decoherence would follow immediately and be succeeded by another coherence, and so on to any length of time. If uniformity in the vigour of the blows could be ensured, each consecutive cycle of operations within the coherer would resemble accurately the preceding one; and the galvanometer would integrate the pulsating currents traversing it and give their mean value. It remained to be seen whether the time allowed between the blows was sufficient for the production of coherence. The trembler first used performed about fifty vibrations per second. The coherer selected was one of moderate sensitiveness. The following figures were immediately obtained by submitting the coherer, while being tapped, to known E.M.F.s taken from a wire potentiometer. Here the galvanometer deflections are left, as obtained, in scale divisions. Fig. 35 is plotted direct from this table.

"TABLE II. (Fig. 35).

Volts.	Deflection in Scale Divisions (1 cm. corresponding to about 3 microamperes).	Volts.	Deflection in Scale Divisions (1 cm. corresponding to about 3 microamperes).
0	0	1.4	1.4
0.1	0	1.5	1.5
0.3	0.1	1.6	1.7
0.4	0.1	1.7	2.4
0.5	0.2	1.9	4.2
0.7	0.4	2.0	5.9
0.9	0.6	2.1	8 to 10
1.0	0.5	2.3	13 to 14
1.3	1.0	2.4	15 to 16

"After the lengthy experience of coherers which had been gained by the writer, the steadiness of the galvanometer reading and the certainty with which increase of deflection followed the movement of the jockey along the potentiometer wire seemed almost incredible. Occasionally, however, there were rather violent throws which sent the spot of light off the scale, and there were also sudden lapses which allowed the spot to fall towards zero. A number of coherers were roughly tried over, and from them figures giving similar curves obtained. It was, therefore, decided

"TABLE III. (Fig. 37).

The letters have the same signification as those in Table II.

Volts.	A	B	C	C′	D	E	F
	Microamperes.						
0	0	...	...	...	...	...	...
0.04	0.3	...	...	...	...	...	...
0.08	1.5	...	...	...	...	...	...
0.1	2.1	...	...	...	...	...	...
0.2	9.0	0.3	...	...	...	...	...
0.3	22.2	0.6	...	...	...	...	...
0.4	31.2	1.2	0.3	...	...	...	...
0.5	45.0	1.8	...	...	...	...	...
0.8	...	3.6	1.2	...	...	...	...
1.0	...	5.7	1.8	...	...	...	...
1.2	...	9.0	2.7	...	...	...	...
1.4	...	14.7	4.2	0.3	...	...	...
1.6	...	21.5	7.5	0.3	...	...	...
1.8	...	...	12.9	0.6	...	...	...
2.0	...	39.0	19.2	1.2	...	...	...
2.2	...	52.2	27.3	1.5	...	...	...
2.4	...	...	40.8	2.1	...	...	...
2.6	...	...	53.4	3.6	...	...	...
2.8	...	...	...	5.8	...	...	...
3.0	...	...	...	11.1	...	...	...
3.2	...	...	...	20.1	0.3	...	...
3.4	...	...	...	31.2	0.3	...	...
3.6	...	...	...	46.2	1.2	...	...
3.8	...	...	...	63.0	1.8	...	...
4.0	...	...	...	...	3.0	0.3	0.0
4.2	...	...	...	...	3.9	...	0.3
4.4	...	...	...	...	6.3	...	...
4.6	...	...	...	...	13.8	...	...
4.8	...	...	...	...	15.9	0.6	...
5.0	...	...	...	...	22.8	1.2	...
5.5	...	...	...	...	55.2	2.7	...
6.0	...	...	...	...	...	8.4	0.6
6.6	...	...	...	...	...	22.5	1.2

to remodel the apparatus and to prosecute thoroughly the new method. It was noticeable that in every case a decided change occurred in the conductivity of the coherer at voltages much lower than the critical voltages obtained by previous processes from the same coherers.

"TABLE IV. (Fig. 38).—MODERATELY SENSITIVE COHERERS UNDER LOW E.M.F.S.

Volts.	Microamperes.		Volts.	Microamperes.	
	1.	2.		1.	2.
0.00	0.00	0.00	0.30	0.27	0.04
0.04	0.03	...	0.44	0.44	0.06
0.08	0.06	...	0.54	0.57	0.09
0.12	...	0.02	0.60	0.68	0.10
0.16	0.13	...	0.70	0.89	0.13
0.20	0.16	0.03	0.80	0.22	0.16

"TABLE V. (Fig. 39).—NON-SENSITIVE COHERERS.

1.		2.	
Volts.	Microamperes.	Volts.	Microamperes.
0	0	0	0
4.0	0.1	6.0	0.3
5.0	0.4	6.4	0.4
5.4	0.7	6.8	0.5
5.8	1.1	7.2	0.7
6.2	1.8	7.6	0.8
6.6	3.0	8.0	1.5
7.0	4.8	8.4	2.2
7.4	7.4	8.8	3.7
7.8	11.1	9.0	5.9
8.2	16.2	...	...
8.6	20.6	...	...
9.0	26.5	...	...

"The apparatus was wholly rearranged. Continuity of sliding contact being of no importance, a potentiometer was made by winding about 100 ft. of bare resistance wire on a long ebonite rod of circular section. When operated by six cells consecutive turns had a difference of potential of 0.02 volt very approximately. The contact piece moved parallel to the axis of the ebonite rod. A moving-coil mirror galvanometer was shunted till its sensitiveness was such that 1 cm. on the scale corresponded to 3 microamperes. A bell-trembler especially capable of adjustment as to stroke and speed was obtained; and a non-inductive resistance of 10,000 ohms was placed in the coherer-galvanometer circuit. Suitable terminals were arranged for the quick interchanging of the coherers, and convenient switches were placed in the various circuits. A diagrammatic plan of the apparatus is given in Fig. 36. With this disposition of the apparatus the characteristic curves were plotted for a large number of tubes, the circumstances of each experiment being maintained for all invariable. A selection of the curves so obtained is given in Fig. 37. These are plotted from Table III., in which the scale readings of the galvanometer deflec-

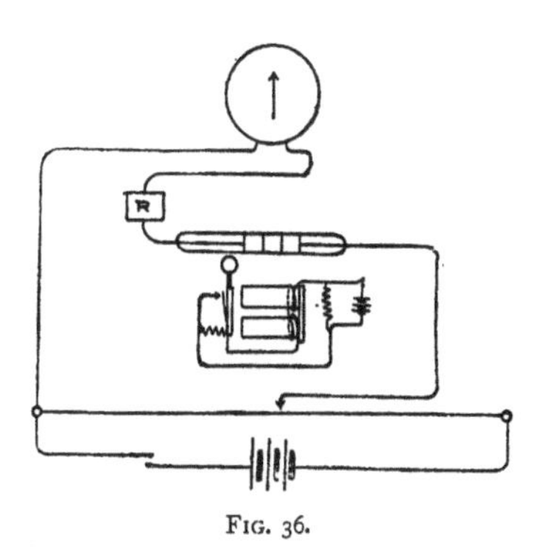

FIG. 36.

tions have been converted to absolute values. From this selection will be seen what was merely emphasised by all the other curves obtained, that the general position of its curve on the chart indicates exactly the sensitiveness of a coherer as judged by the radiation test. It should be noticed that the occurrence of rapid coherence is on these curves indicated by their convexity towards the axis of E.M.F.—were the conductivity of the coherer constant the curves would degenerate into sloping straight lines passing through the origin. With all the coherers examined, excepting those of known and remarkable non-sensitiveness, some curvature was apparent under comparatively small E.M.F.s. Further, there was in no case, as the applied E.M.F. was increased, any sudden increase in the current; but, on the contrary, always a gradual growth in its value. Again, it will be noticed that when the applied E.M.F becomes great enough the relation between current and E.M.F. becomes approximately linear, indicating that

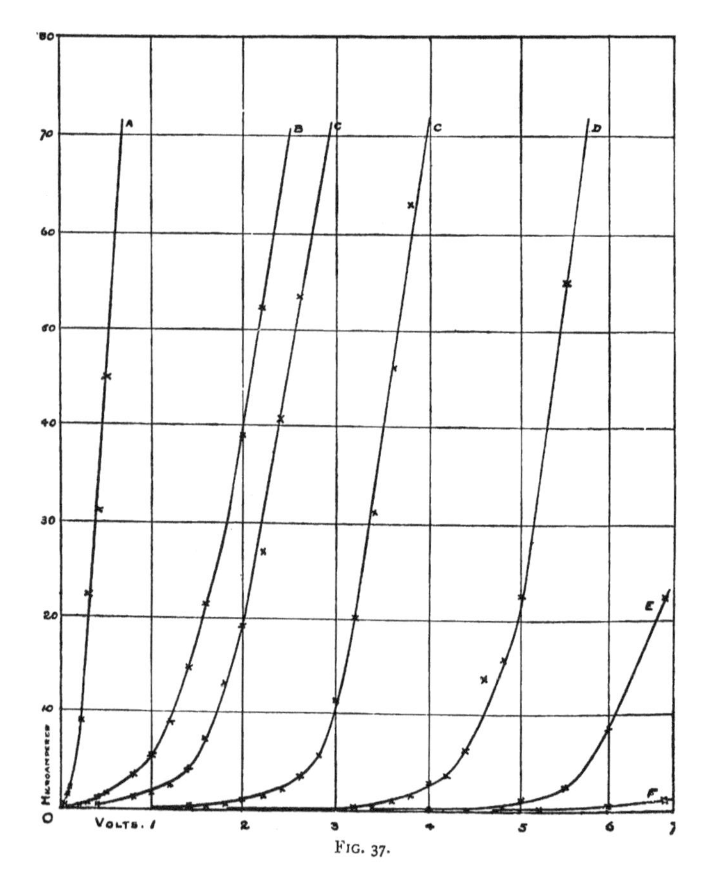

Fig. 37.

coherence is now increasing uniformly and slowly. In order to examine more closely whether coherence could be traced at rather low voltages the sensitiveness of the galvanometer was raised by changing the shunt, and two of the coherers whose curves are given in Fig. 37 submitted to low E.M.F.s, the frequency of the trembler being unaltered. Fig. 38 (Table IV.) gives the curves so obtained. They seem to indicate that the fall of resistance of a coherer commences very early, but, also, that no great change-rate is reached till a certain E.M.F. is applied which is different for each coherer. Another and similar trial on some very non-sensitive tubes, in which P.D.s up to 9 volts were used, showed that the linear portion of the curve for most coherers could, if desired, be reached. Two of these

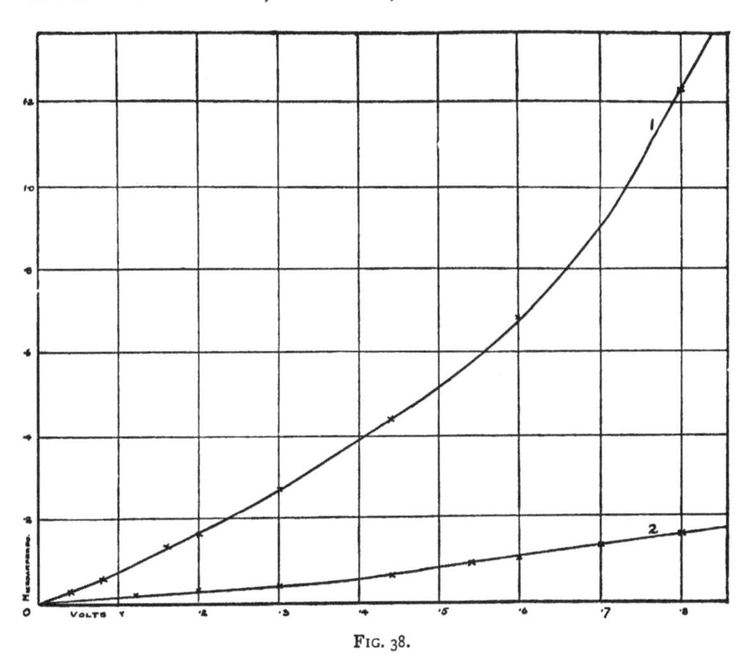

FIG. 38.

curves, which are in every feature similar to those in Fig. 37, are given in Fig. 39.

"In the late summer opportunity arose of applying this method of examination to a very large number of coherers. But before beginning the work an attempt was made to overcome certain irregularities. For instance, it had been noticed that the same tube tested again and again gave curves which were not quite coincident, and sometimes the motion of the galvanometer mirror was erratic. These and other small vagaries were traced to variations in the speed of the vibrator, to variations in the vigour of the blow, and, not least, to the occasional occurrence of minute sparks at the contacts of the vibrator. In the absence of a convenient mechanical tapper, the bell-trembler used was supplied with a small-resistance shunt across the spark-gap, and means were taken to steady the current energising the magnets. Several sizes of hammer were tried, and the weight of the blow and the speed of tapping altered through a considerable range. No such

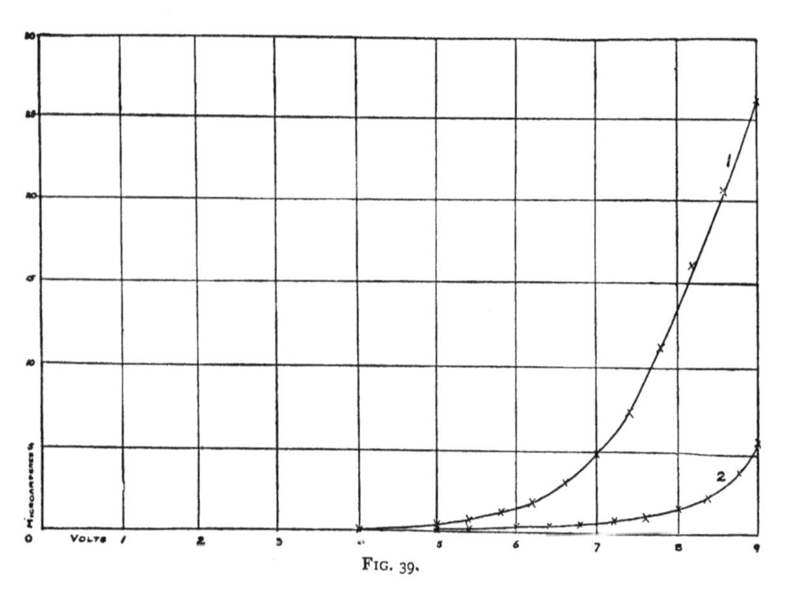

Fig. 39.

alteration changed the character of the curves, so convenient conditions were adopted and adhered to. The speed of tapping chosen had, judging by ear, a frequency of about fifty vibrations per second. It was arranged that the inductance of the coherer circuit was as low as possible.

"With these precautions several hundreds of coherers were examined. The curves obtained were more regular than those obtained hitherto, but presented no new features. The occasional throws so noticeable while the spark-gap of the trembler was only slightly shunted were comparatively infrequent. The final conclusion which these trials led to need here be merely stated. The complete correspondence between the sensitiveness of a coherer to electromagnetic radiation and the position on the chart of the characteristic curve obtained by this method was fully demonstrated.

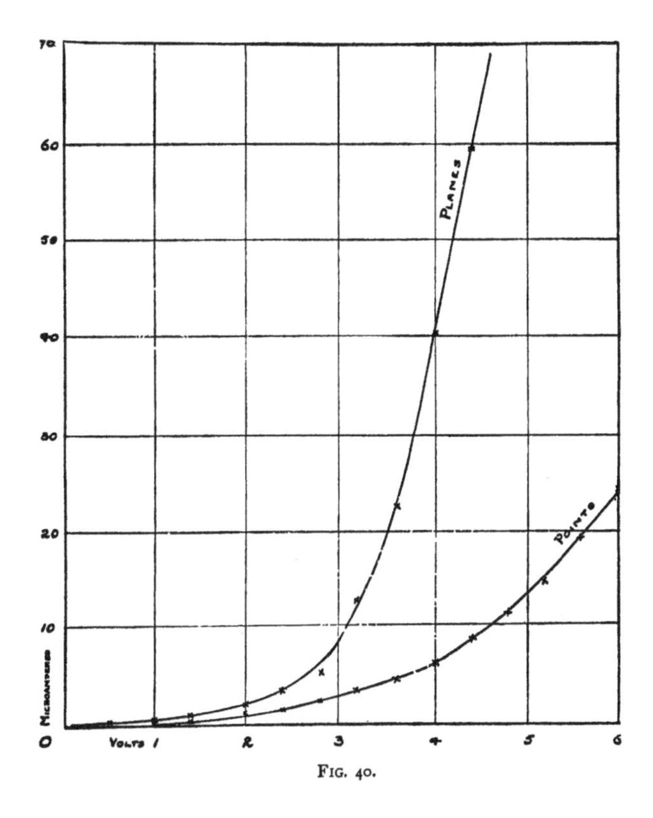

FIG. 40.

"It may here be mentioned that the majority of the coherers examined were of one type, consisting, as before stated, of filings, usually nickel, between plane silver electrodes. The author purposely kept to this one type for a reason which will be seen later. Other types of coherers were, however, occasionally examined. As a matter of interest there is given in Fig. 40 an example of the curves obtained from a coherer which consisted of two steel needles projecting co-linearly from opposite directions into iron filings, the needle points being about 1 mm. apart. In the same figure is also a curve given by a coherer of the ordinary type, but containing iron filings. The general results of the mass of observations briefly described in the preceding pages, and a study of the characteristic curves, led the author to a kinetic theory of the filings coherer, which makes the typical coherer consist of filings free to move throughout the space of a narrow crevasse bounded at its ends by plane conducting faces. It is necessary to state this hypothesis tersely, so that the aim of the experiments to be described may be understood.

"TABLE VI. (Fig. 40).—COHERERS WITH IRON FILINGS.

Volts.	Microamperes.		Volts.	Microamperes.	
	Planes.	Points.		Planes.	Points.
0	0	0	3.6	22.8	4.8
0.5	0.3	0	4.0	40.5	6.3
1.0	0.6	0.3	4.4	59.4	8.7
1.4	1.2	0.6	4.8	...	11.1
2.0	2.1	1.5	5.2	...	14.4
2.4	3.6	1.8	5.6	...	19.2
2.8	6.6	2.7	6.0	...	24.3
3.2	12.6	3.3			

"When a difference of potential, steady or oscillatory, is established between two opposed plane conductors separated by a distance which is small compared with their own magnitude, a fairly uniform electric field is created between them. Any conducting piece of matter, not spherical in shape, moving freely in this field, will experience a couple tending to place it, in general, with its longest axis parallel to the field.* Other particles present will be similarly acted upon, and if they come to rest while the electric field is maintained, they may be expected to exhibit distinct

* This much, the author believes, has been suggested before, but he has not been able to find any distinct statement.—W. H. E.

orientation. The degree of orientation among the mass of filings will evidently bear some relation to the strength of the electric field, that is, to the applied E.M.F. In this state the process of 'coherence' will be easier than when the arrangement of the particles is fortuitous. If we take Dr Lodge's view, that an actual welding occurs between contiguous particles till complete chains are formed across the gap, it must be noticed that the perfection of the weldings will also, in all probability, depend on the E.M.F. producing them. Thus, in filings coherers, the writer thinks, two distinct actions—one, the mechanical, resulting in orientation of the particles, the other, the true 'coherence' phenomenon, resulting in weldings—occur, and these are both dependent on the applied E.M.F. It is the author's contention that the share of the mechanical action is great. The series of experiments whose description follows were designed to prove the existence and the importance of this mechanical action.

"In seeking for modes of attack, the idea of sorting filings according to their size and shape presented itself prominently. A coherer containing only spherical particles, and others, each containing uniform ellipsoidal particles of selected degrees of elongation, ought, in the ideal case, to be constructed and compared. In an endeavour to realise these conditions, the author spent much time in sifting silver, aluminium, iron, nickel, and copper filings. But even after many siftings all these filings, on examination under a microscope, were of astonishing irregularity as regards both size and shape. A process of sorting by allowing sifted filings to fall through liquids was then tried, but proved equally unsatisfactory. Finally, after many other attempts—which included such methods as rolling the filings down a vibrated inclined glass plate, and the use of air blasts—it was decided to abandon this phase of the work, inasmuch as the use of filings not perfectly regular could not but yield inconclusive results. It is known that if the gap between the electrodes of a coherer be increased, the sensitiveness of the coherer is diminished. It is mentioned later that a certain constant in the theoretical equation of the curves is directly proportional to the length of the gap. This constant is, roughly, for each position of the electrodes, that value of the abscissa at which rapid curvature commences. A number of experiments were therefore performed on several coherers of the ordinary type. First, the characteristic curve of a coherer was obtained by the general method described, and the distance between the plugs measured. Then the tube was heated towards the part where the platinum wire emerges, and, by a strong pull, the electrodes separated slightly. When the tube had cooled, another characteristic curve was drawn, and the gap again measured. This operation, it was found, could only be repeated once, or at most twice, on home-made coherers. The results of a series of measurements conducted in this manner on one tube are given diagrammatically in Fig. 41, the measurements from which the curves are plotted being set out in Table VII. A glance at the last lines of the table will show the reader that the accordance between the crude theory and the experimental data obtained is fair, but not remarkable enough to call for emphasis.

"TABLE VII. (Fig. 41).—EFFECT OF INCREASING THE GAP.

Volts.	Microamperes.			
	1	2	3	4
0.	0	0	0	0
0.1	0.3	0	0	...
1.6	0.9	0.3	0.3	0
2.0	1.8	0.6	0.3	0.3
2.4	6.0	0.9	0.9	0.6
2.6	14.2	1.5	1.5	0.6
3.0	42.0	3.0	2.1	0.9
3.4	...	7.5	4.5	1.5
3.8	...	22.2	10.2	3.0
4.2	...	37.8	21.3	7.5
4.6	...	...	34.5	16.5
5.0	...	...	58.0	34.2
5.4	...	...	...	61.0
Curvature commences at Size of gap	$E=2.1$ $l=0.22$	$E=3.0$ $l=0.34$	$E=3.3$ $l=0.38$	$E=3.5$ volts. $l=0.45$ mm.
	$E/l=9.5$	$E/l=8.8$	$E/l=8.9$	$E/l=7.8$

'Only a day or two before going to press it occurred to the writer that Tissot's discovery, that the sensitiveness of a coherer made with magnetic filings is greatly increased by the presence of a magnetic field whose direction is parallel to the axis of the tube, might be utilised to indicate the share the mechanical actions have in the alterations of

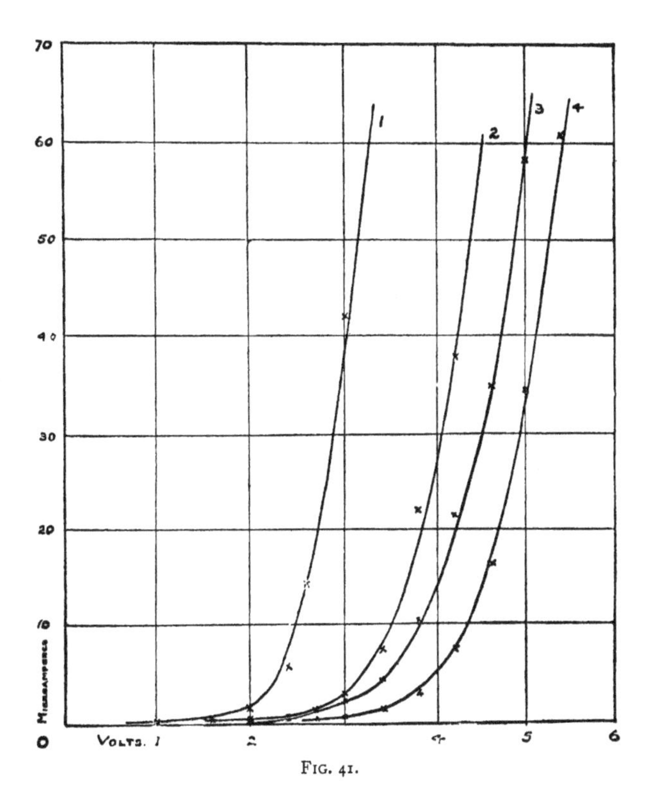

Fig. 41.

conductivity in a filings coherer. Magnetic particles in such a field will experience a couple similar in kind to that which they experience in an electric field. A short calculation shows, besides, that the couples exerted on the same (paramagnetic) particles by an electric field due to 1 volt across a small gap and by a magnetic field of one C.G.S. unit are of the same order of magnitude. A knowledge of the permeability of the material of the filings is however, necessary to make any experimental results valuable. The writer has not yet had opportunity to perform by the characteristic curve method any quantitative experiments with iron filings coherers, but he has found that the effect of the magnetic field due to a small solenoid is quite appreciable on a nickel-alloy coherer which happened to be at hand. Tissot states that the effect of a magnetic field on the sensitiveness of a coherer containing iron filings is very great.

"*Elementary Theory of the Motion of the Particles.*

"If a difference of potential of e volts be established between the parallel faces of the electrodes of a filings coherer, and if the distance between the faces be l cm., the approximately uniform electric field in the narrow gap will be of intensity $F = e/l$. This field intensity is independent of the specific inductive capacity of the medium occupying the gap, and, therefore, if we assume that the presence of the flying metallic particles is merely equivalent to an increase of the specific inductive capacity, the above expression for the field intensity will be unaltered by the presence of the filings. An application of the well-known formula for the couple on a solid prolate spheroid in a uniform magnetic field shows that in an electric field of strength F, the couple tending to bring the long axis of the ellipsoid towards coincidence with the direction of the field is given by the expression—

$$4\pi F^2 \sin\theta \cos\theta \frac{M-L}{LM},$$

where θ is the angle between the major axis and the direction of the field, and L and M are constants which are somewhat complicated functions of the axes of the ellipsoid. The particle will therefore acquire in its flight an oscillatory motion the same in nature as that of a quadrantal pendulum (Thomson and Tait, Part I. p. 392). The period of small oscillations can in any particular case easily be computed. For example, if we take a particle whose length is 0.02 mm. and two or three times its minor axis, of mass $\frac{1}{2} \times 10^{-4}$ gramme, moving in a field due to a

P.D. of 1 volt across a gap of 0.2 mm., we find that its time of swing, when the oscillations are small, is of the order $\frac{1}{5000}$th of a second. Thus the time required to pull a particle into line with the field is very small indeed if its angular displacement from that line is small. In the actual case of coherers subjected to the experimental processes described in the preceding part of this paper, the filings in the gap were, by blows repeated at intervals of $\frac{1}{50}$th of a second, scattered throughout the space between the electrodes. Observation shows that they have time in each cycle to settle at the bottom of the space free to them. They settle while under the influence of an electric field. Very many more of the particles, therefore, will fall with their long axes parallel to the axis of the tube than would so fall had no field influenced them. Moreover, the more will so fall the greater the couple acting on each particle—that is, the greater the field, or the greater the applied E.M.F.

"If we agree with those who affirm that chains are formed (see Lodge's letter to *The Electrician*, vol. xlv. p. 938) among the mass of filings during coherence and assume that the number of complete chains so formed is related linearly to the number of particles brought to rest all parallel, we may, for our rough purposes, deduce an expression for the conductivity of the coherer by calculating the proportion of the flying particles which will be pulled into parallelism with the axis by a given electric field. It is to be remembered, however, that in all probability the cohesion or welding between particle and particle is itself dependent on the strength of the field, that is, on the electromotive force applied at the electrodes; but this question it is not intended to discuss. Besides which, there are also other and less abstruse considerations which bear on the chance of a complete chain being formed. This chance is the greater, for instance, the greater the length of a particle and the smaller the distance across the gap.*

"To avoid going into details it may be stated that by making suitable approximations a simple form of the equation of the characteristic curve is easily arrived at. The

* Previously noticed by Malagoli.

equation reduces, in fact, to the hyperbola

$$y = \frac{x - C}{R + \dfrac{B}{x - E}}$$

where y is the current through the coherer when x is the

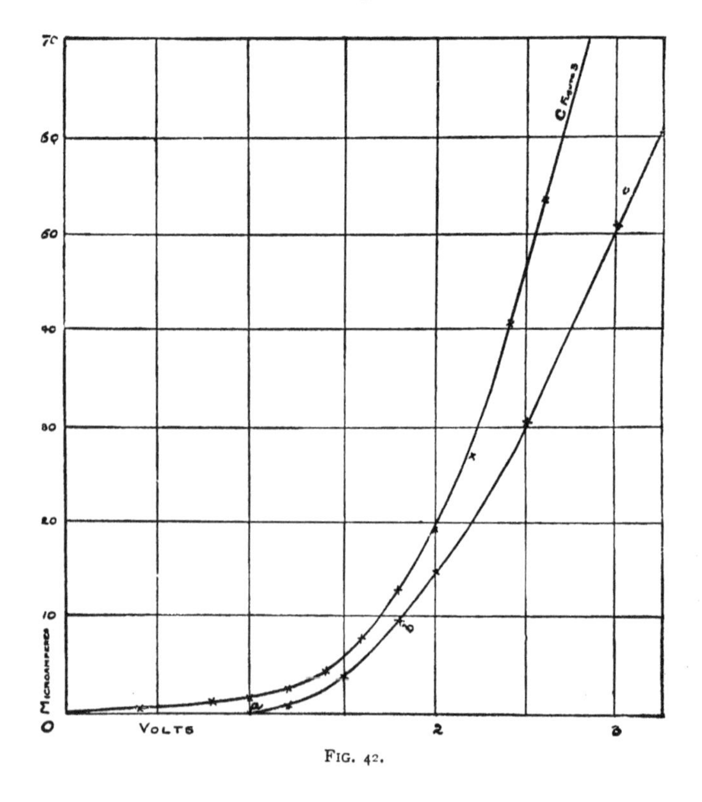

FIG. 42.

applied E.M.F., R is the external resistance, B and C are constants, and E is a constant which is proportional to the length of the gap and which it is well to keep prominently in view. Only a portion of one branch of this hyperbola is applicable to the present case, and that, moreover, only while the amount of coherence is not great. This portion represents, besides, but a part, though the most important part, of the whole characteristic curve of a coherer, but it enables us to compare the whole action of the filings coherer with that part of the whole action which can possibly be due to the motion of the particles.

"TABLE VIII. (Fig. 42).—EXAMPLE OF CALCULATED CURVE.

x (volts).	y (microamperes).	x (volts).	y (microamperes).
1.0	0	2.0	14.7
1.2	0.7	2.5	30.6
1.5	4.1	3.0	50.7
1.8	9.7		

E=1.0. C=0.17. B=60.

"It will be noticed that this approximate equation makes the current zero when the applied E.M.F. is E. Thus this constant E is that E.M.F. which just suffices to turn particles only very slightly displaced into complete coincidence with the axis, and of these there is theoretically only an infinitely small proportion of the whole number of filings. In other words, it is that voltage at which the characteristic curve of a coherer begins to bend sharply. The bending which occurs before that E.M.F. is reached is due to the coherence which may be expected to take place in the actual case at some of the many contacts which are afforded even when the filings fall perfectly fortuitously. These views are excellently expressed by the curve in Fig. 42, which has been selected for this purpose. The line *abc* represents the calculated curve, E being taken as 1 volt, B as 60 kilohms, and C as 0.17 volt. It indicates how much of the whole action might be effected by the mechanical agencies alone. But there is reason to believe, from considerations not to be entered upon here, that the constant C has in this case been taken unduly large, and that as it stands the curve somewhat exaggerates the share of the mechanical action. The amount by which the calculated curve falls short of the experimental curve affords, however, an idea of the proportion of induced conductivity for which coherence proper is responsible. The writer has found, by using different resistances in series with the coherer, that this degree of coherence proper is the greater the smaller the external resistance, that is, the larger the current, and thus that the coherence proper depends on the amount of current allowed to pass through the coherer. Since only small currents were permitted in obtaining the curves given in this article, the true coherence effect was of less weight than the mechanical effect.

"*Conclusion.*

"So much has been said on the main points at issue as they arose during the progress of the descriptive parts of this article, that a lengthy discussion of any of these points seems now quite unnecessary. But it would be well to state as concisely as possible the general tendency of the writer's work on the whole subject of coherers made with filings. The first series of experiments with the liquid and wire potentiometers demonstrated satisfactorily that electrical surgings, such as are no doubt produced in any circuit to which an E.M.F. is abruptly applied, had no effect—as surgings—in producing coherence. From this result alone it may be concluded fairly that, as Branly and others have always insisted, the phenomenon of coherence is an effect of E.M.F. merely. But, besides, these first experiments brought prominently before us the many vagaries which have caused the filings coherer to be universally regarded as an instrument of erratic and unaccountable behaviour.

"The next portion of the work led to the invention of a very general and powerful method of examining filings coherers. By the use of this method all the anomalies which have hitherto accompanied the manipulation of a coherer seem to disappear, and a curve is obtained which expresses most definitely all the characters of the coherer from which it is drawn. It may be termed a true quantitative method of investigating the qualities of a coherer. As one example of its utility in such investigations the curves in Fig. 41, tracing the effect of separating the electrodes, may be cited. The curves indicate plainly that very definite and invariable laws govern the action of a coherer. They prove, besides, that coherence is a continuous phenomenon, and that there can be no true 'critical voltage.' Moreover, the complete agreement which the author found, during his work with hundreds of coherers, between the sensitiveness to electro-magnetic radiation and the position on the chart of the curves of the respective coherers, prove most emphatically that the ultimate causes of the whole phenomenon of coherence are the same with oscillatory and with non-oscillatory E.M.F.s. Thus those common expressions, 'the impinging of waves,' and 'the incidence of radiation on a coherer,' might, with advantage, the writer thinks, be deleted from our phraseology. For the production of coherence by radiation is in all probability entirely due to oscillations of potential set up in the coherer connections.

"The theory which the writer has put forward to explain the merely mechanical portions of the complex phenomenon of coherence in filings coherers is, the writer recognises, not yet backed by any irrefutable evidence. Against it can be brought many objections. One which will occur to every reader will take the form of the question: Are there sufficient complete oscillations in the radiation which falls, ordinarily, on a coherer circuit to produce an appreciable motion in one of the particles? As has been mentioned above, an average filing of the dimensions which the writer found by rough measurement under a microscope required not more than $\frac{1}{20000}$th of a second to turn from rest to parallelism with the electric field acting upon it. For the couple to be maintained so long a time about 100 oscillations of the period commonly employed in wireless telegraphy would be necessary; and this is not impossible in resonant circuits. On the other hand, the proposed theory throws light on many points hitherto unexplained. For instance, the ability of tremors, and even of sounds, to assist coherence and the increase of sensitiveness which follows a proper application of a magnetic field to coherers containing magnetic filings become more easily understood. And, again, the 'critical voltage' of Blondel and Dobkevitch is shown to have a cause though its existence as an experimental actuality is made impossible by the simultaneous occurrence of coherence proper. The theory shows very clearly, too, how liable to yield conflicting conclusions must be all comparisons made by different observers on coherers supposed exactly similar, and how doubtful in value are comparisons made, even by the one experimenter, on coherers containing filings of different metals. For, the behaviour of different instruments must depend very largely on the size and shape of the filings. That highly contradictory results have been obtained by different workers was pointed out in the preliminary part of the article. Finally, if the theory be accepted, the direction of improvement in the manufacture of coherers is, by its aid, most clearly indicated. On the whole the writer thinks the aim of this article may well be described as an attempt to reduce the complicated case of the filings coherer to the elementary case of the single microphonic contact."

The results of Dr Eccles' admirable research may be summarised in a few words:—(1) The filings coherer acts on the application of a difference of potential to its terminals, and this voltage may be either continuous or alternating. (2) Any small voltage is sufficient to produce some current, but in the neighbourhood of a certain voltage, which is different for each coherer, the resistance decreases very rapidly, allowing a correspondingly rapid increase of current, which welds the filings, and causes a further decrease of resistance. (3) The sensibilities of coherers for steady voltages are proportional to their sensibilities for oscillatory voltages. (4) The whole phenomenon may be explained mathematically on the supposition that the action takes place in two stages—firstly, the turning of the particles into line under the influence of the electrostatic force, and secondly, a welding together of their ends, thus brought together, owing to the passage of a current.

Lieutenant Solari, of the Italian Navy, discovered that a small drop of mercury between two iron plugs forms a very sensitive self-restoring coherer. Its disadvantage lies in its need for frequent adjustment.

Dr Muirhead has improved upon this by using a small rotating disc of steel which grazes the surface of a pool of mercury covered with a film of oil. The continual motion of the disc prevents mercury from sticking to the steel, and at the same time ensures that the surface of the mercury is clean. One advantage of this coherer is its low resistance when cohered, which makes it possible to use it in connection with a syphon recorder or other sensitive telegraph instrument without interposing a relay.

Anti-coherers.—If the effect of electrical oscillations is to increase the resistance of a detector, the apparatus is called an anti-coherer. The most important of these are perhaps the electrolytic device used by Dr de Forest, and Mr Brown's lead peroxide detector, both described in a later chapter. Unlike the metallic contact coherer, which may be cohered by the electric jigs whether the local battery is connected or not, the anti-coherer is only maintained in its sensitive state by the action of a steady voltage in its circuit, and is restored to its original state by the same agency after the reception of a signal. In the case of the coherer, mechanical vibrations perform a similar function. Carbon coherers are, under ordinary circumstances, self-restoring. It seems likely, however, in default of experimental evidence to the contrary that the restoration may be due to mechanical vibration, which enables the contact again to attain a more or less stable state of equilibrium.

Schafer's Anti-coherer.—One of the most curious of the anti-coherers is that invented by Schäfer, and tested at distances up to 95 kilometres in Germany. A fine scratch is made with a diamond across a strip of silver deposited chemically on glass. One end of the strip is connected to one pole of a battery giving 3 or 4 volts, and the other end to the other pole. Under these circumstances, a galvanometer in circuit shows quite an appreciable current, and a microscope reveals the fact that the small torn particles of silver in the gap are flying to and fro, from side to side, acting as carriers of a convection current as in the well-known electrical experiment usually called

"Electric Hail." If the two parts of the strip be in addition connected to the aerial and earth wires of a receiving station, and have thus a voltage jig impressed upon them, the motion stops, and the current is no longer carried. In practical use this change of current is detected by means of a telephone.

In seeking for an explanation of this action we must note that the gap is not filled with metallic particles as is the case in a coherer, but that the particles are comparatively few in number, and have very large areas in comparison to their weights, since they are made from an extremely thin metal sheet. We must also take into account that the electric field between the two sides

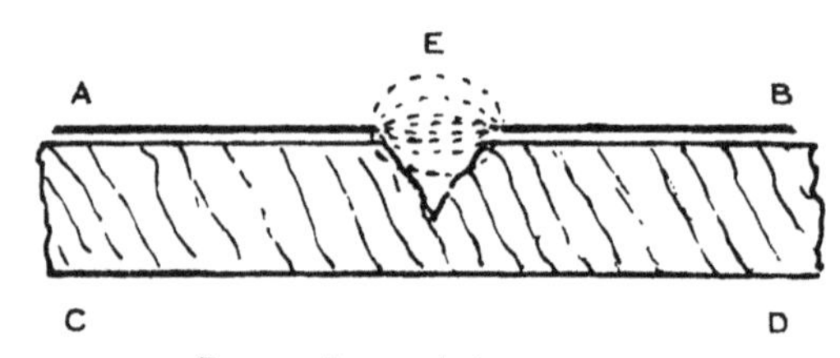

FIG. 43.—SCHAFER'S ANTI-COHERER.
A B, Silver Film; *C D*, Glass Plate; *E*, Lines of Electric Force across Gap in Film.
(The diagram is roughly 1,000 times natural size.)

of the gap is not uniform, since the thickness of the silver film is much less than the width of the gap. The lines of force are therefore closer together near the sides than at the centre of the gap.

The action of the particles in carrying the current under the influence of a steady voltage is simple enough. An uncharged particle near the electrode is attracted, owing to the difference of potential. It approaches until it comes into electrical contact, and then a charge flows into it from the electrode. When the electrical surface density on the particle reaches a certain limit, which depends on the potential at the electrode and on the electrical dimensions of the particle, the particle is repelled from the electrode and flies across to the other side and there gives up its charge.

So far we have dealt with what is already well known. I shall now apply these facts to the elucidation of Schäfer's detector. Consider what must happen to a charged particle, which has just commenced its flight, if the potential of the electrode is rapidly changed, as it would be by the arrival of the first half wave of a jig. When the potential of the electrode rises considerably, the sign of the charge remaining the same, an opposite charge will be induced on the near side of the particle. As the field is divergent, the effect will be to reduce the repulsion between electrode and particle. If the potential and charge on the electrode now diminish to zero before the particle has moved appreciably farther, attraction will ensue since the particle retains its original charge. This attraction will be further increased as the second half of the wave changes the sign of the electrification on the electrode, which now becomes oppositely charged to the particle. On the whole, then, the action of a change of voltage at the electrodes, superposed upon the steady voltage which charged the particles, is to reduce the repulsion or cause attraction, and, in general, to stop their flight.

These considerations appear to give a sufficient explanation of the phenomenon observed by Schäfer, but I may as well restate them in a somewhat different form, which is briefer, thus:—An uncharged particle is attracted, during both positive and negative halves of an oscillation, to the nearest electrode; a charged particle, beginning its flight under a steady voltage, is repelled from the nearest electrode. If the two conditions be superposed they tend to balance one another and arrest the movement of the particle. The action is not molecular, and therefore does not resemble that of other anti-coherers.

Messrs Duddell and Taylor's Measurements of Voltage in the Coherer Circuit.—With the kind permission of the authors, and of the Council of the Institute, I append a quotation from the paper,* read to the Institution of Electrical Engineers in 1905, in which the authors give an estimate from experimental data of the actual voltage in the coherer circuit :—

"Though no precise measurements bearing on the point were made, it may be taken in view of the manner in which the coherer receiver used at intervals during the experiments responded at the longer distances, that the E.M.F. of 6.6 volts given in the previous paragraph was about eight times stronger than necessary to get signals on that particular coherer. The critical R.M.S. electromotive force during each train in the receiver antenna required to operate the coherer, would therefore be about $\frac{6.6}{8} =$ 0.8 volt. This was not the actual voltage applied to the coherer terminals, however, because the coherer was always connected in a secondary circuit tuned to resonance, in which it may be taken that the E.M.F. was magnified, perhaps ten times, giving 8.0 volts per train (R.M.S.). In this secondary resonant circuit a special form of construction has been adopted, and every precaution taken to reduce the damping to a minimum. To get an idea of the maximum voltage attained during a train of oscillations in this circuit, it is necessary to take into consideration the increment and decrement of the train. Allowing for this, it is thought safe to assume that an actual maximum E.M.F. of about four or five times the R.M.S. figure, viz., about 40 volts, should be available in the coherer resonant circuit to give good signals. As in practical work it would be necessary to provide a considerable margin above the figure, it is clear that the maximum voltages produced are not of a very small order."

* *Journal I.E.E.* vol. 35, No. 174.

CHAPTER V.

DETECTION OF OSCILLATORY CURRENTS OF HIGH FREQUENCY BY THEIR EFFECTS ON MAGNETISED IRON.

THE effect of an oscillatory magnetic field on magnetised steel or iron differs considerably in different circumstances. A permanent change, however, of the state of magnetisation of the iron is almost always apparent. The results depend on the qualities and magnetisation of the material, on the intensity and character of the oscillations, and on the relative directions of the magnetisation and the oscillatory field. It has been known since the earlier part of last century that the discharge of a Leyden jar will magnetise or demagnetise steel needles. Professor Joseph Henry made a large number of observations on the phenomena, and threw much light on the nature of the discharge of a condenser, but it was not until 1895 that the possibility of detecting, by this means, electric waves which had travelled considerable distances was proved by Mr Rutherford. In that year, by using a large Hertzian oscillator with plates about 1 by 2 m. and a resonator consisting of two copper rods about 30 cm. long, he was able to detect the radiations at distances up to three-quarters of a mile. The rods of the resonator were connected by a small helix of fine wire, in the axis of which was placed a magnetised needle or bunch of very fine steel wires. A magnetometer indicated the amount of magnetism in the needle, and decrease of its deflection showed the arrival of the waves. After each observation it was necessary to remagnetise the needle, no automatic arrangement for this purpose being provided.

Professor Rutherford suggested to me in 1900 that the oscillations might be made to record themselves on a steel band, as in the telegraphone. I was, however, too fully employed with other work at the time to undertake the experimental work necessary to produce a practical wireless telegraph receiver.

In 1897 Professor E. Wilson constructed an automatic magnetic receiver. It was similar to Rutherford's first apparatus, but the remagnetisation of the needle was effected by a local battery whose circuit was closed by the motion of the magnetometer. This appears to be the first *telegraphic* receiver in which the magnetic detector was employed.

Five years later, in 1903, Mr Marconi patented two forms of magnetic detector somewhat different in principle from those already described. Instead of allowing the steel to remain demagnetised for an appreciable time and then remagnetising it by closing a local circuit, he applied a slowly alternating field to the iron core. As the core goes through its cycle of magnetisation, the latter lags behind the applied force. If, however, an oscillatory field of high frequency is suddenly superimposed on the slowly alternating one, the molecules of the iron are shaken up and the lag instantaneously disappears, the magnetisation jumping at once into the condition corresponding to the applied field. A similar effect is produced by a mechanical shock. The sudden change in induction through the core is observed by means of a telephone receiver in circuit with a coil of wire round the core of the detector. On the arrival of oscillatory currents at the detector, the momentary current induced by the change of magnetic induction causes a click in the telephone.

Mr Marconi's second magnetic detector attains a similar end in a somewhat different way. An endless band of fine insulated iron wires is kept moving at about 8 cm. per second in the direction of its length round two wooden pulleys. Near to one part of the band two horse-shoe magnets are fixed, two similar poles being together, and the opposite ones further along the band on either side. A large number of lines of force pass into the band under the central poles, and after travelling some distance in either direction along it reach the outer poles of the horse-shoes. Wound on a glass tube, through which the band

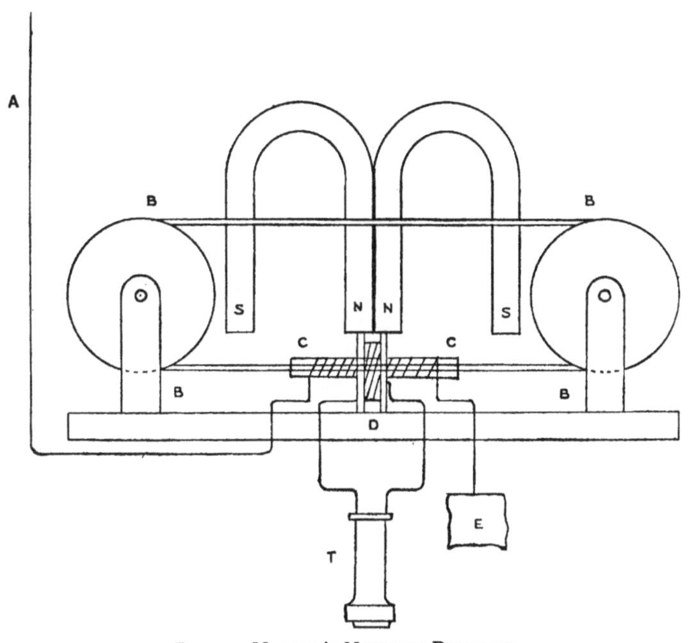

FIG. 44.—MARCONI'S MAGNETIC DETECTOR.

A, Aerial Wire; *E*, Earth-Plate; *B*, *B*..., Iron Band round Pulleys; *S N*..., Permanent Magnets; *C C*, Primary Winding on Glass Tube through which the Iron Band Travels; *D*. Secondary Winding; *T*, Telephone Receiver.

passes at this point, are two coils of fine wire. The inner one is connected to the aerial and earth wires, and the outer to the terminals of a telephone receiver. As the band travels slowly along, the lines of force are drawn out in the direction of its motion owing to the retentive property of the iron, a proportion of those which cut one side of the telephone circuit when the band is at rest are, therefore, drawn away from it. When the oscillatory currents arrive from the aerial circuit these lines are freed and fly back, cutting one side of the coil as they go (see Russell, p. 111, "Retentiveness"). The result is a momentary current which produces a click in the telephone.

Each spark at the transmitter thus makes an audible sound in the receiver. As in other forms of wireless telegraph a short succession of sparks constitute a dot, and a long one a dash. Messages in the Morse code are, therefore, transmissible. This receiver is more sensitive than the coherer, and is also more easily adjusted. It has come into very general use on account of these advantages, but its inability to actuate a recording instrument is in certain circumstances a drawback, particularly if the operator be not extremely expert and the message be in cypher. In Post Office work over the ordinary wires the difficulty is obviated by the use of an automatic type-printing telegraph where possible, thus eliminating errors which might occur at the receiving end. In other cases the repetition of the message, back from office to office, is insisted on by the Post Office as essential to accuracy, when cypher or code words occur in it.

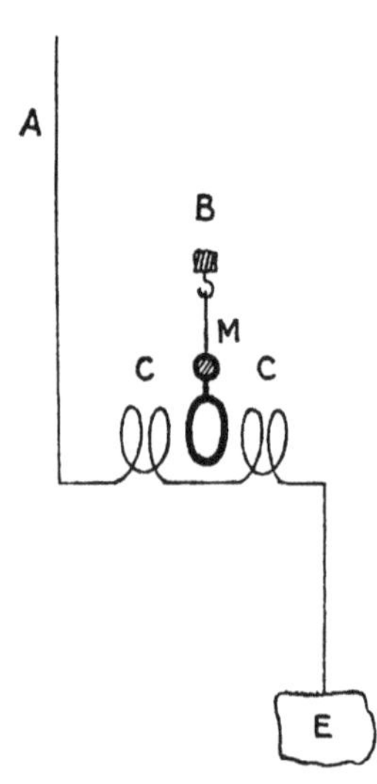

FIG. 44A.—FESSENDEN'S ELECTRO-MAGNETIC RECEIVER.

A, Aerial Wire; *B*, Support for Moving Parts; *M*, Small Mirror, below which is a Light Ring of Wire.

In 1902 Professor Fessenden invented several forms of magnetic receiver. The earliest depends on the reaction between a closed circuit of low resistance and another circuit, or coil, in which oscillatory currents are flowing. The closed circuit is a very light ring of wire to which a mirror is attached, the whole being suspended by a fine fibre. The oscillatory currents flow through a fixed coil, inducing currents in the ring which repel those in the fixed coil. The signals are read by the deflections of a spot of light from the mirror, as is frequently done in submarine cable work. No iron or steel was used in this instrument.

His second detector, shown in Fig. 44B, is entirely different in principle, and depends on the magnetic properties of steel. A fine steel wire is included in the line between the aerial and earth wires, with its length at right angles to a powerful magnetic field. One end is also connected to a local battery circuit, the other pole of which forms a terminal, very close to, but not in contact with the steel wire. The wire is normally cross magnetised by induction from the magnetic field, and is therefore subject to a force tending to pull it into the position where the field has greatest density; it thus sags in the direction of the poles of the fixed magnet. The oscillatory currents along the wire suddenly reduce the permeability, or cause some other change which is equivalent in effect, and the wire springs up and strikes the contact, completing the local circuit. The use of a recording instrument in the local circuit seems possible with this detector, though the telephone would no doubt be more sensitive. On the whole the apparatus does not appear suitable for the detection of very minute currents.

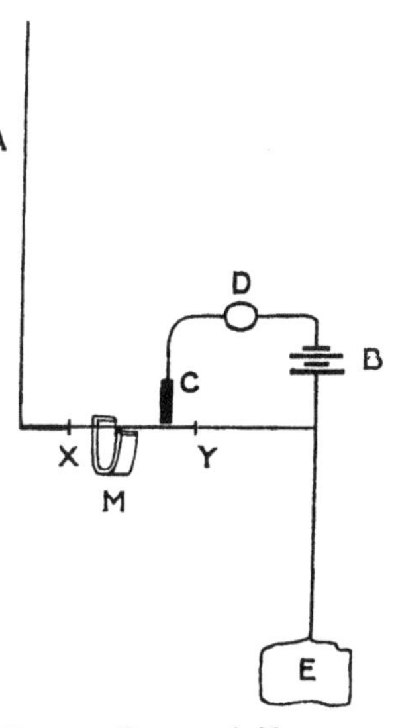

FIG. 44B.—FESSENDEN'S MAGNETIC RECEIVER.

A, Aerial Wire; *X Y*, Thin Steel Wire; *E*, Earth; *M*, Permanent Magnet; *C*, Fixed Contact Piece; *B*, Battery; *D*, Galvanometer.

In the recently invented "Einthoven" galvanometer

which is extraordinarily sensitive to continuous currents, a single wire in a very strong magnetic field is also used, but the wire is of non-magnetic material; with oscillatory currents it would therefore tend to vibrate in time with the oscillations at a frequency too great to be observable, even if the forces concerned were sufficient to cause a mechanical vibration of more than infinitesimal amplitude; the adaptation of this instrument to the needs of wireless telegraphy seems therefore unlikely.

In 1904 Mr Walter and Professor Ewing patented a magnetic detector based on the discovery that oscillatory currents *along* a fine steel wire increase its hysteresis. A description of this apparatus was given in a communication to the Royal Society in 1904,* from which the following is an excerpt:—

"It occurred to us to exhibit the alteration in hysteresis by applying the principle which is used in an instrument invented some years ago by one of us for the mechanical measurement of hysteresis. In that instrument† the hysteresis is measured by the mechanical couple between a magnetic field and the iron, when either the iron or the magnet providing the field is caused to revolve. Thus, if the field revolves, the iron tends to be dragged after it, as a consequence of hysteresis in the reversals of its magnetism, and if the motion is prevented by a spring or other control, it assumes a deflected position. Suppose, now, the electric oscillations to act on it, any change of the hysteresis caused by them will be exhibited by a corresponding change in the deflection. We anticipated, in accordance with the generally accepted view that hysteresis is reduced by the oscillations, that their presence would be detected by a fall in the deflection.

"With this expectation an experimental apparatus was arranged, consisting of an electro-magnet, capable of being rotated on a vertical axis by an electric motor. The magnet poles were bored out circular, and between them was sus-

* *Proc. Roy. Soc.* vol. 73, February 1904, by kind permission of the Council.

† *Journ. Inst. Electr. Engineers*, vol. xxiv. pp. 398-430, 1895.

pended, by a phosphor-bronze strip, a ring made up of three thin, flat annuli of soft iron, clamped together, and provided at the foot with an axial pivot. The ring was free to turn inside of two bobbins wound with fine copper wire, the windings being at right angles to the plane of the ring. Through these copper windings, electrical oscillations, produced in the usual manner by means of a distinct spark-gap, were passed.

"The first experiments resulted in a very small deflection from the position due to normal hysteresis, indicating, as was expected, a decrease of hysteresis when the oscillations arrived. The apparatus was also tested with an alternating current of about 100 periods per second in place of the oscillations, with the effect that the normal hysteresis deflection was almost entirely wiped out.

"Various other forms were also tried with indifferent results, when it occurred to us that there would be advantages in passing the oscillations through the magnetic material itself, making it of magnetic wire. A small bobbin was therefore wound with insulated soft iron wire, and the ends soldered to the upper and lower halves of the spindle, which was itself divided at the centre, the upper half bearing the controlling spring, and the lower dipping into mercury, from which a connection led to the other terminal. On passing oscillations through this winding, a remarkable and unexpected result was obtained. The change of deflection was much more marked than in the former experiments, and was in the opposite sense, indicating an increase of hysteresis while oscillations were present. Afterwards, hard steel wire was substituted for the soft iron, and a very great increase in the effect was observed, still in the same direction—that of increase of hysteresis.

"Owing to these encouraging results, it was decided to continue the experiments in this direction, abandoning the older form, in which a decrease of hysteresis was dealt with. The first bobbin constructed was about $\frac{5}{16}$ inch in external diameter, and had a vertical wire space of $\frac{1}{4}$ inch. The winding was a single No. 32-gauge iron wire, double cotton-covered, wound straight round from beginning to end. Later, No. 40 and No. 46 steel wires were employed, of which the latter gave the best results.

"It was soon noticed that any method of increasing the oscillatory current in the wires, as by winding the bobbin with two wires having a slightly unequal number of turns, was of advantage in giving a larger deflection. Later a fine copper wire secondary, wound on the bobbin parallel

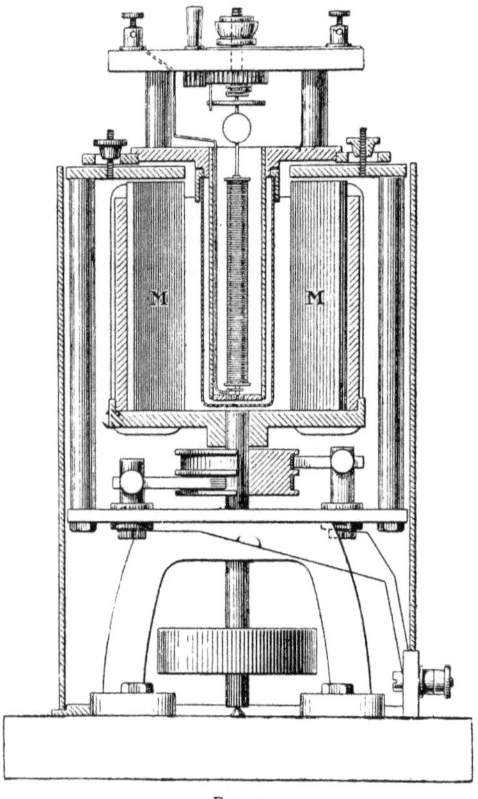

FIG. 45.

to the magnetic wire, was tried, first with the ends insulated, and then with the ends soldered together. A marked increase in deflection was observed when the secondary was closed, showing that the magnetic nature of the wire itself was influential. Accordingly, a bobbin was then wound with insulated steel wire, doubled back on itself. This non-inductive winding gave by far the best results hitherto attained, and is now used, except when special results are required.

"The instrument, though described as a detector of electrical oscillations, may be said to measure rather than detect, giving quantitative as well as qualitative results, and being capable of regulation from a sensibility of the same order as that of an average coherer down to practical insensibility to powerful sparks in the same room.

"In the instrument, as shown in the figure, the electro-magnet takes the form of a ring capable of moving round a vertical axis, and is provided on the interior with two long wedge-shaped pole-pieces, M, M, the current to the winding being supplied through brushes bearing against insulated rings below. The magnet is made to revolve by an electro-motor, the best speed being about five to eight revolutions per second, but the electro-magnet may be replaced by a permanent magnet system giving a similar field. A structure is built up, external to the magnet, to support the vessel containing the pivoted bobbin and its centring arrangements. The bobbin itself is made of bone, and is about two inches long. It is provided with a steel spindle at each end bearing in a jewel hole, the two halves of the spindle being insulated from one another. The winding, which is, as far as possible, non-inductive, consists of about 500 turns of No. 46-gauge hard-drawn steel wire, insulated with silk. The bobbin is immersed in petroleum, or a mixture of petroleum with thicker mineral oil, which serves the double purpose of fortifying the insulation, and giving the damping effect necessary to steady the deflection due to the drag of the revolving magnet. Readings are taken by means of a spot of light, as with speaking mirror galvanometers, but a syphon-recording attachment has been fitted, and any form of contact for working a relay could be employed.

"The detector, as before mentioned, gives quantitative readings, and, in some cases, the deflection may be too large to be easily read by the scale. For this purpose a variable shunt is provided, by which the deflection can be regulated.

"For the purpose of wireless telegraphy, the instrument has the advantage of giving metrical effects. The benefit of this in facilitating tuning, and in other respects, need not be insisted upon."

Mr Walter has more recently invented another form of self-restoring magnetic detector based on a quite different principle and capable of producing unidirectional—that is, direct—currents when excited by the received oscillations. This forms so great an advance on previous detectors that with his kind permission I quote his description of the apparatus, given in a paper to the Royal Society in April 1906: *—

"The method was arrived at as a result of experiments in connection with an instrument previously described,† to determine the cause of the increase of hysteresis loss as a result of the action of oscillations. It was found that the increase is due to a great extent if not entirely to the increase of induction produced, to which increased induction a largely augmented hysteresis loss corresponds at the field strength employed. Working on this basis, it was thought that such an increase of induction might serve as a means of furnishing continuous unidirectional currents, by generating an unidirectional (commuted) E.M.F., *i.e.*, by making conductors cut the lines of force in a magnetic field, and causing the oscillations to act upon a magnetic mass undergoing reversals of magnetism in the magnetic field of the generator, whereby the E.M.F. generated should be augmented; a second, equal E.M.F. being opposed to the first, so that normally there is no external potential difference. In such a case a continuous unidirectional current should be obtainable during the time that the oscillations are acting upon the magnetic mass.

"An experimental apparatus was accordingly made, a diagrammatic plan of which is given in Fig. 46. Two ebonite bobbins, B, B_1, mounted on the same spindle, are rotated in the field of two horse-shoe permanent magnets, NS, N_1S_1, these bobbins being wound, in a similar manner

---

* *Proc. Roy. Soc.* A. vol. 77, April 1906, by kind permission of the Council.

† Walter and Ewing, *Roy. Soc. Proc.* vol. 73, p. 120, 1904.

to those illustrated in connection with the pivoted bobbin detector previously referred to, with some feet of steel wire of suitable resistance. A winding of two coils, W, W′, at right angles to one another, of a hundred turns, is placed on each bobbin, at right angles to the plane of the steel wire winding, as in a drum armature, corresponding coils, *i.e.*, W and W_1, W′ and W'_1, being connected in such a way that the E.M.F.'s generated are equal and opposite. The ends of the windings are connected to the segments of a 4-part commutator, C. (For the sake of clearness only one pair of corresponding windings, of one turn each, is shown connected in Fig. 46.) The steel wire windings of the two bobbins are exactly alike, the ends of one winding being insulated, while those of the other are connected to a pair of slip-rings, *r*, *r*, and brushes, by means of which the oscillations can be passed through the winding.

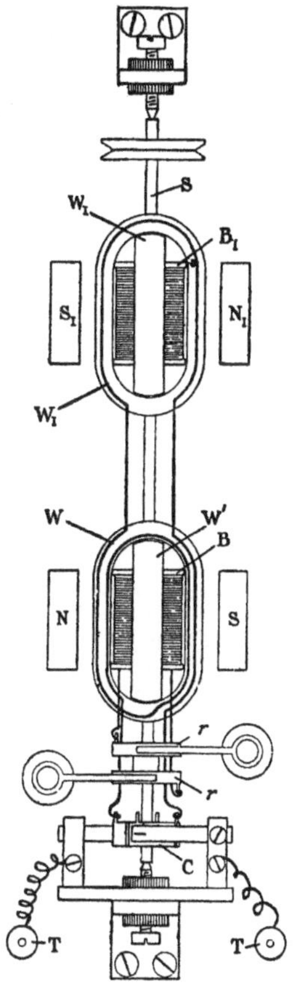

Fig. 46.

"On testing this apparatus in the normal condition, with the armature driven by a small electric motor, and no oscillations acting, there was no potential difference at the brushes, the zero of a sensitive Ayrton-Mather galvanometer connected to the terminals, T, T, remaining undisturbed. On waves arriving, a steady deflection on the galvanometer was obtained, in a direction corresponding to an increase of E.M.F. generated by the armature (bobbin) acted upon by the oscillations. On the oscillations ceasing the galvanometer deflection returned to zero. The effect naturally was very small in the first experiments, but it has been found that by suitably designing the magnetic winding and proportioning the turns in the armature winding a quite considerable sensibility is obtained, and this is continually being improved upon. The usual speed employed is about five to eight revolutions per second; higher speeds have been tried and give a larger effect, but the zero is not so steady.

"The model illustrated is not adapted to give the best results, this form having been chosen solely for convenience in construction. A considerable length of the winding on the armature is "dead" wire, and hence in a new model being constructed, the armatures resemble small Gramme ring structures, in which the wire is more effectively utilised.

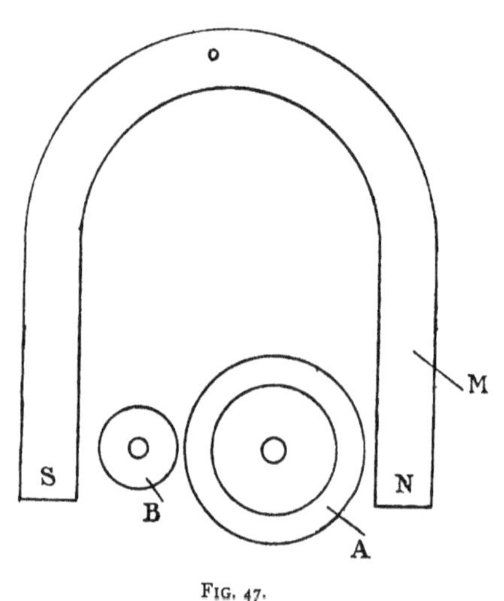

FIG. 47.

"The results obtained with the first form of the apparatus led to the idea that the magnetic mass might be located elsewhere in the magnetic circuit of M, such as at B in Fig. 47, undergoing slow continuous reversals at the most favourable speed, and an ordinary ring armature, A be used, which latter could then be run at a much higher speed so that a proportionately greater external potential difference as a result of oscillations acting could be anticipated, two identical generators, opposed to one another of course, being employed as in the previous method. The few experiments made in this direction have, however, not given good results up to the present, but this is considered to be due rather to the experimental apparatus employed than to the inapplicability of the method."

The paper by Mr Russell, referred to by Mr Walter, was read before the Royal Society of Edinburgh in 1905, and is the description and discussion of a series of experiments on the subject. The conditions of the experiments differed somewhat from those of Messrs Ewing and Walter, or of Dr Eccles, whose most recent research on the subject I shall quote further on. Mr Russell used a closed secondary circuit of an induction coil linked with a closed magnetic circuit of thin sheet iron, to which a steady field was also applied. As there was no spark-gap it is unlikely that any *high* frequency oscillations took place, the secondary current being simply of the type shown in Fig. 32. Thus the results may or may not be directly comparable with those obtained with magnetic detectors in wireless telegraphy.

Mr Russell has kindly permitted me to make the following quotations from his paper,* noting that the results given are not only true for electrical oscillations but are also observed with a mechanical vibration :—

"It is now absolutely necessary to discriminate between the order and manner in which oscillations and field are superposed, the one upon the other. Two experimental methods were adopted :—

"*A*. Oscillations were superposed upon constant field.

"*B*. A change of field was superposed upon oscillations permanently acting.

"*Experimental Methods under A Conditions.*

"First, after demagnetisation of the iron, a fixed maximum field is put on (by increasing reversals, to secure as far as possible symmetry about the zero of induction), and reversed twenty times. The plus change of inductions due to the twenty-first reversal is measured. The co-directional oscillations are now superposed, and the plus induction change measured. Oscillations and field are now put off. Second, the iron is again demagnetised, and the same fixed

---

* *Proceedings R.S.E.* vol. xxvi. part i.; by kind permission of the Council.

maximum field put on in the same way as before, and reversed twenty times. The plus induction change due to the twenty-first reversal is measured. A single step is now taken to any given point on the hysteresis loop, and the minus change of induction measured. The co-directional oscillations are now superposed, and the plus or minus reading taken. Oscillations and field are again put off. This second process is repeated for a sufficient number of points all round the loop. The whole process is repeated for transverse oscillations.

"Three curves result from the reduced galvanometer readings taken as above described. These are plotted in Figs. 48 I. and 48 II., when the maximum cyclic induction values are (without oscillations) B=780 and B=5,620 respectively. The scale of Fig. 48 II. is for both ordinates double that of Fig. 48 I. The ordinates measuring induction are in C.G.S. units; the abscissæ measuring field, in arbitrary units.

"*Summary of Results under A Conditions.*

"At and near extreme cyclic values the superposition of oscillations produces for low values of field (see Fig. 48 I.) a relatively large increase of induction; for higher values of field (see Fig. 48 II.), a relatively small increase of induction. For low values of field the increase is greater for co-directional than for transverse oscillations; for high values of field this relationship is reversed.

"After leaving cyclic extremes there are points when the field is decreasing where co-directional and transverse oscillations respectively produce neither an increase nor a decrease of induction. In low fields they are thrust *from* the cyclic extremes, in high fields *towards* the cyclic extremes.

"When these points are passed, oscillations produce a decrease of induction, and this for all values of field is greater for co-directional than for transverse oscillations. This decrease passes into increase in the opposite sense, and the first conditions are reverted to at the other end of the cycle.

"In all cases the induction change is greatest when oscillations are superposed on an increasing field. For low fields this occurs at or near *cyclic extremes*, where the slope of the curves is greatest. But as the cyclic field maximum is increased, the greatest induction change occurs at an *earlier stage* of the increasing field, where in this case also the curves are *steepest.*

"The curves for transverse and co-directional oscillations given in Figs. 48 and 49 must not be confounded with the

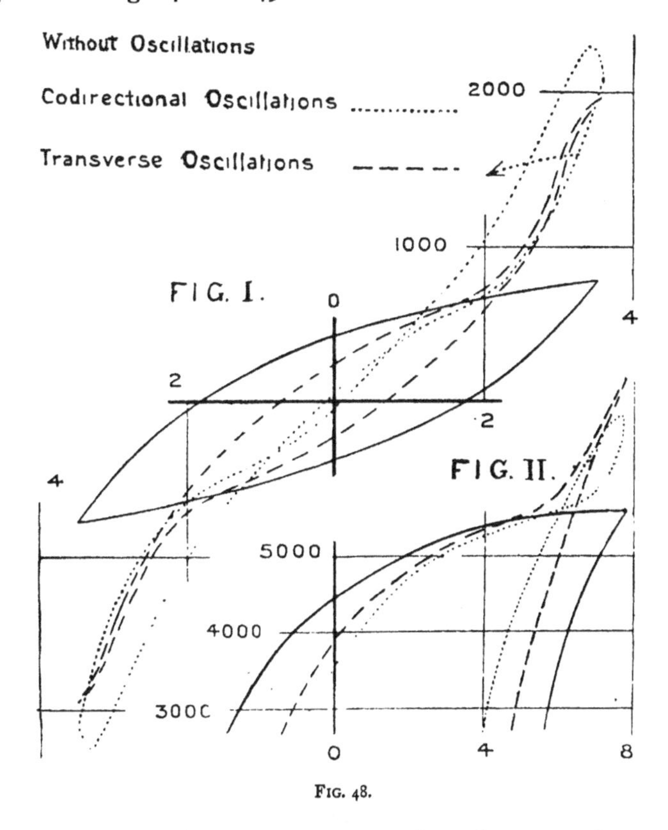

FIG. 48.

usual hysteresis loops in the sense that the areas they enclose measure the energy loss during one complete cycle. They do not do so.

"They measure for any given value of field the instantaneous change of induction which takes place when oscillations—co-directional and transverse—are superposed at any and all stages of the normal hysteresis loop.

"Now suppose that after any instantaneous induction change has been measured, the field is made to vary by some small amount—say, by a decrement if the field had previously been decreasing—the induction change which now takes place is entirely different (see dotted arrow). Hysteresis or lag in the usual sense comes into full play, and one naturally passes to the conditions of field superposition where a cycle field change may be regarded as superposed upon permanently acting oscillations.

"*Experimental Methods under B Conditions.*

"1. After demagnetisation and twenty reversals of a fixed maximum field, the normal B – H hysteresis loop is determined by Ewing's method of single steps from the fixed maximum to a sufficient number of points all round the loop.

"2. The iron is again demagnetised and subjected to co-directional oscillations, upon which the field at the same fixed maximum value is superposed. After twenty reversals of field, the hysteresis loop is determined as before.

"3. The iron being again demagnetised, the same process is repeated for transverse oscillations and the corresponding measurements made. It is, of course, understood that in cases *second* and *third* the force sustaining the oscillations remains 'on' and unaltered until the series of galvanometer readings has been completed.

"The above determinations were repeated for many field cycles, the maximum induction values at the extremes of each cycle ranging from a minimum of B = 20 to B = 12,000.

"*Summary of Results under B Conditions.*

"*Permeability.**—See Fig. 49, where the full line, dotted line, and dash line curves have the same signification as in Figs. 48 I. and 48 II. For low values of field co-directional oscillations increase the permeability relative to the normal (*i.e.*, without oscillations) to a greater extent than transverse oscillations.

"For higher values of field transverse oscillations, in-

* After twenty reversals of field.

crease the permeability relative to the normal to a greater extent than co-directional oscillations.

"The crossing point of these curves occurs when the induction is about 5,000 with oscillations.

"When the induction does not exceed a few hundreds without oscillations the corresponding induction with transverse and co-directional oscillations is respectively about three or four times greater. When B = 20 without oscillations, these ratios become a little less.

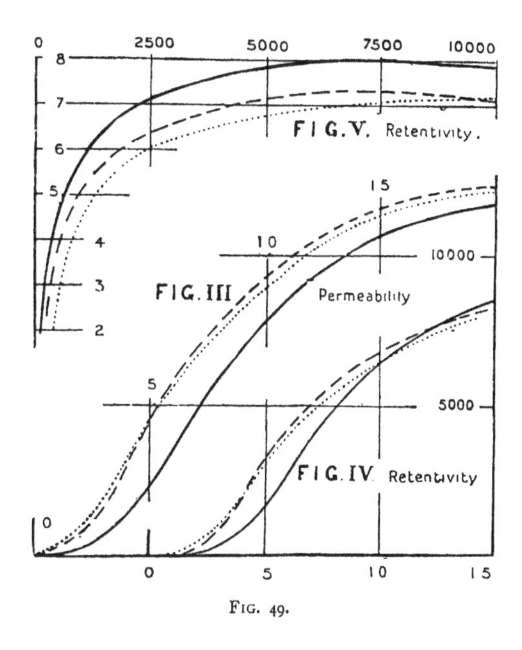

Fig. 49.

"*Retentivity.*—When the field is reduced to zero from a cyclic maximum and the ratio of residual to maximum induction plotted, as in Fig. 49, against maximum cyclic induction, the curve with transverse oscillations is higher within wide limits than that obtained with co-directional oscillations. These curves appear to coalesce, or even to cross, in higher fields. (The normal retentivity curve lies above both, not applicable for 'quenched' nickel.)

"It might thus appear that even under the B conditions the effect of oscillations is to reduce residual magnetisation.

Such a statement, however, cannot be regarded as correct, because by plotting against maximum induction one of the most important effects of oscillations has been eliminated, viz., that of increased permeability.

"In Fig. 48 residual magnetisation is plotted against field, and we see that it is only at high values of induction—where permeability is, so to speak, naturally eliminated—that the effect of oscillations is to reduce retentivity. For lower fields, and throughout a wide range, oscillations *increase* retentivity. A comparison with Fig. 49 shows that the greater the permeability in the three cases, the greater is the residual magnetisation. The residual magnetisation curves with co-directional and transverse oscillations also cross each other—as was found to be the case for permeability—and greatly exceed in value the normal retentivity curve without oscillations, with the exception already mentioned, where the induction is great.

"*Coercive Force.*—For low fields of the order of a few hundreds, oscillations likewise increase coercive force, but with slightly higher fields this effect soon disappears, and thereafter oscillations decrease coercive force.

"*Hysteresis Loss for Constant Induction.*—When under normal conditions the maximum cyclic induction is of the order of one to five hundred, co-directional oscillations for constant maximum induction diminish the energy loss in the iron about four times, transverse oscillations about three times. But as the induction is increased this difference gradually lessens, and after B under normal conditions has reached 5,000 it is apparent that transverse oscillations diminish the loss in the iron to a greater extent than co-directional oscillations. In all cases for constant induction, oscillations cause a diminution of the energy loss; but when the induction is high (say, B = 12,000), the diminution, although sufficiently well marked, has become a relatively small effect.

"*Hysteresis Loss for Constant Field.*—When under normal conditions the maximum cyclic induction is of the order of a few hundreds, oscillations for constant maximum fields *increase* the energy loss in the iron about four times relative to the loss when no oscillations are acting. As the induction is increased to some thousands, the energy loss becomes very approximately the same with and without oscillations.

When, however, the induction is higher still (say, $B = 12{,}000$), a sufficiently well-marked but relatively small decrease of hysteresis loss is caused by the oscillations. The energy loss for co-directional and transverse oscillations does not differ greatly relative to each other for constant maximum fields. The greater relative retentivity and the lower permeability at low fields (where the difference is so much greater than under normal conditions), under transverse relative to co-directional oscillations are in harmony with this result. The curves exhibited are not here reproduced. One however may be referred to, in which the full line curve shows the normal hysteresis loop without oscillations, the dotted line curve, the greatly increased hysteresis loop with co-directional oscillations. They do not differ in type from each other. The phenomenon of 'lag' is equally well exhibited by both; but for present purposes it is in my opinion also essential to state *all* the facts in terms of permeability (at cyclic extremes), of retentivity (when $H = 0$), and of coercive force (when $B = 0$), as has been done above."

Dr W. H. Eccles has recently investigated the phenomenon under conditions which approximate very closely to those obtaining in actual wireless telegraphy. He has kindly permitted me to make the following excerpt from his paper: *—

"In the experiments to be described, an endeavour has been made to turn the difficulty arising from the skin-effect. Oscillations so feeble have been used that they affected only the outermost layers of the iron wires employed. The cores of the iron wires have, therefore, not been used.

"Other and great difficulties arise in the matter of producing oscillations of determinate and invariable character. Maurain, in the greater part of his work on this subject, appears to have used the oscillations that passed through a helix in series with a Leyden jar kept sparking strongly and continuously by means of an induction-coil. Russell,† in some recent experiments, applied to his iron the oscillations passing through a coil connected directly

* By kind permission of the Council of the Physical Society and of the publishers of the *Philosophical Magazine.*

† *Proc. Roy. Soc. of Edin.* Nov. 20, 1905.

in series with a small induction-coil. This last method appears to the writer to subject the iron to very violent treatment of a nature not easily described accurately; for how far the mere surgings of secondary current overwhelm in importance the genuine oscillations, it will be difficult to say. Piola,* again, worked with highly damped oscillations, because, he found, such oscillations produce the highest effects. He has, indeed, following Rutherford, used this fact in determining the damping factor of a circuit. But in the present investigation these sources of indeterminateness have, as far as might be, been avoided by using a single train of waves instead of a continued torrent of such trains, and by using oscillations as little damped as possible.

"The present paper deals solely with oscillations whose magnetic field is along the direction of the principal field.

"In these experiments fairly soft Swedish charcoal iron, aged, and not freshly annealed, was taken a large number of times round any chosen magnetic cycle (as in ordinary magnetic testing) till what may be called a cyclic state was attained. The field was then given any desired value, and the iron submitted to a single train of oscillations. This was managed by generating the oscillations on a helix surrounding the iron wire. The consequent alteration in pole strength was observed by the deflexion of a magnetometer mirror. These processes were all repeated a number of times for each selected point of the cyclic curve. The figures given in the table below are thus each the mean of a number of observations.

"In detail, the apparatus adopted consisted of two straight solenoids each of 3,270 turns of No. 20 copper wire wound in six layers on a length 59 cm. of split brass tube. They were placed horizontally on one and the same magnetic east-west line, but on either side of a mirror magnetometer. The magnetometer-needle was in the common axis of the two solenoids. The solenoids were connected in series, and so adjusted in position that when a large current was passed through them the magnetometer-needle was undisturbed. A liquid resistance, a Weston milliammeter, suitable switches and commutators, and a battery of six secondary cells completed the solenoid circuit. The

---

* *Elettricista*, iv. p. 145, May 15, 1905.

coil destined to be the seat of the oscillations consisted of 1,252 turns of No. 26 copper wire wound in a single layer on 70 cm. of a glass tube 0.5 cm. in external diameter, and had a resistance of about 3 ohms. The whole coil was wrapped in paraffined paper and pushed into the brass tube of the east solenoid.

"The iron wire examined was unannealed Swedish charcoal iron, diameter 0.749 mm., and was used always in lengths of 56 cm. Its characteristic curve is given in Fig. 50. The magnetometer was a silk-suspended mirror carrying four very small magnets; readings were taken on a scale distant 88 cm.

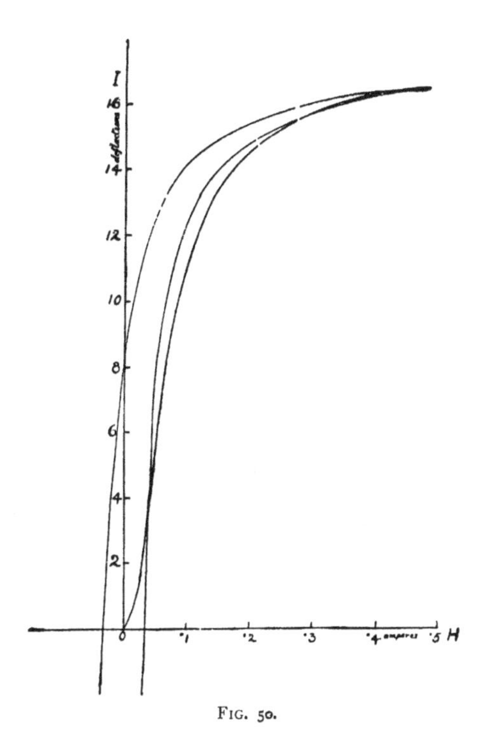

FIG. 50.

"The object in employing twin solenoids is clear. By placing equal amounts of iron wire in each solenoid and adjusting their positions carefully, the magnetometer deflection could be kept very slight whatever magnetic variations the iron was taken through, the iron in the one solenoid compensating that in the other.

"It was then permissible to exalt greatly the sensibility of the magnetometer. This was done by reducing the controlling field at the needle by means of an auxiliary permanent magnet.

"In the experiments, three iron wires insulated from one another and tied in a bundle were used in each solenoid, the group in the east coil lying, of course, in the glass tube on which the oscillation-coil was wound. The inner ends of the wires were 9.3 cm. distant from the magnetometer-needle. The field at the needle was reduced to 0.034 C.G.S. unit. With these arrangements, such a sensibility was attained that on certain days of August last the strokes of distant lightning, not visible in London,* were easily perceptible in the laboratory by their effects on the iron in the oscillation-coil.

"The oscillations used were produced by leading the free east end of the oscillation-coil, that is the end distant from the magnetometer, to one side of a micrometer spark-gap. The other end of the oscillation-coil was left insulated. The poles of the spark-gap were connected to the terminals of a diminutive influence machine, that pole not connected to the oscillation-coil being, besides, earthed. Thus, when the handle of the influence machine was turned through a certain angle (depending on the spark-gap) at a speed easily learned, a spark occurred which set up oscillations in the coil. The spark-length finally settled upon was about half a millimetre. It was found possible in this way to get over and over again practically the same magnetometer deflexion for every spark, provided the effect of previous oscillations was wiped out by taking the iron through a cycle. As these small sparks were usually inaudible, their occurrence was recognised by the sudden deflexion of the magnetometer-needle.

"The calculated period of the stationary waves on the oscillation-coil is 5.7×10^{-7} second.

"A typical set of operations was as follows. The iron was demagnetized by reversals. An amplitude of cycle being then decided upon, the iron was taken a number of times through that cycle. This last operation was stopped at the point settled upon for the observations, and the magnetometer reading taken. A spark was passed, and the magnetometer again read after thirty seconds. A complete cycle was now performed ending at the same point as before, and another spark passed, and so on. Usually, about twenty points on a cycle were examined by 4 or 6 observations at each. In this paper, however, the effects of

* The newspapers gave accounts of thunderstorms in Hertfordshire.

the spark are recorded as if the observations had been taken only on the ascending half of the hysteresis curve; the figures given being in fact the means of the measured effects at points symmetrical with regard to the origin on the ascending and descending branches.

TABLE I.

Current in Milliamperes.	−150	−100	−75	−50	−40	−30	−20	0	20	30	40	50	75	100	150	200
50 cycle	...	...	...	...	0.06	0.10	0.13	0.28	0.38	0.45	0.60	0.55	...	...	...	...
100 cycle	...	...	...	0.27	...	...	...	0.68	1.08	...	...	1.26	1.30	1.18	...	...
150 cycle	...	0.20	...	0.44	...	...	...	0.92	...	1.32	...	1.51	1.62	1.57	1.19	...
200 cycle	0.20	0.36	...	0.68	...	...	...	1.22	...	1.57	...	1.71	1.75	1.55	0.93	0.38

"In this table the figures given can be reduced to absolute measure as follows:—The numbers at the top of each column and down the left column, representing the current through the solenoids in milliamperes, yield when

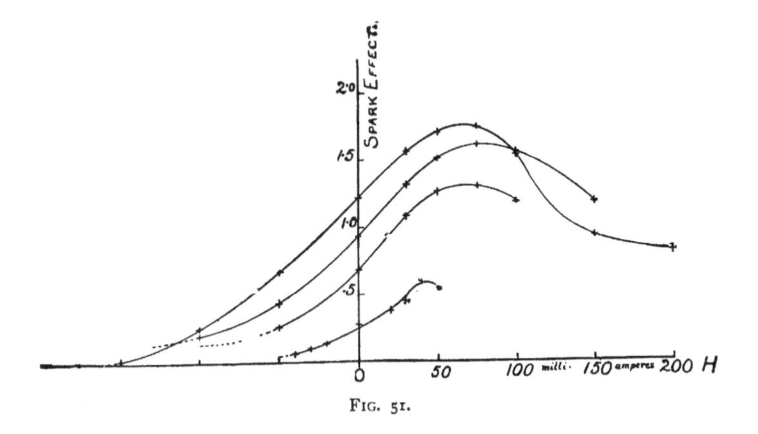

FIG. 51.

multiplied by 0.0696 the magnetic field in C.G.S. units applied to the iron. The deflexions in the body of the table give the change in pole strength, due to the spark, by multiplying by 0.017; yield the volume-average change of intensity of magnetization by the factor 1.27; or give change of total magnetic moment of the affected specimen by the reducing factor 0.940.

"Fig. 51 is plotted from the above table. The curves show clearly how for increasing cyclic amplitudes, within the range here attempted, the effect of the same spark is increased. The curves show very distinct maxima. It is evident, moreover, that the magnitude of the effect at any point is closely connected with the slope of the hysteresis curve. On the whole the curves tend to corroborate the

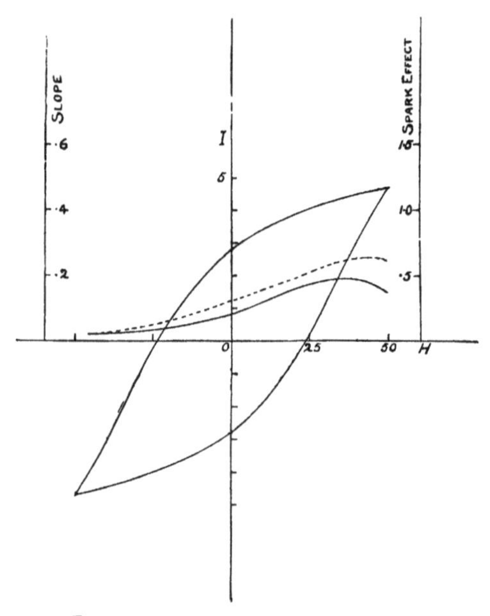

Fig. 52. The dotted curve shows spark-effect.

fact—first noted by E. Wilson in his 1902 patent specification—that the sensibility of the iron to oscillations is greatest when in the magnetic condition represented by the point of inflexion on the hysteresis curve. To examine this matter more closely these curves representing the effect of the spark are repeated in Figs. 52, 53, 54, 55, each alongside the gradient curve, that is, first derived curve, of the corresponding cyclic curve. To assist in picturing the magnetic state of the iron, there is plotted on each diagram the proper hysteresis loop. These loops were obtained by independent experiments with the earth's field controlling the magnetometer-needle. (Reducing factor for I is 67.5.)

"In these figures the abscissæ for all the curves are milliamperes (multiply by 0.0696 to get C.G.S. field). The figures have been so constructed that, except in the cyclic curves, equal ordinates on the different figures represent absolutely equal effects. The reducing factors for the

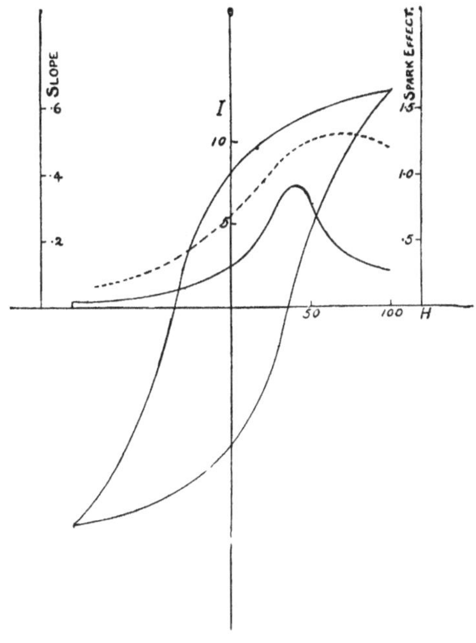

FIG. 53. The dotted curve shows spark-effect.

spark-effect curves are as given in connection with Table I. In all the slope curves the ordinates give at once very approximately the change of intensity of magnetization per .001 C.G.S. increment of field. But the ordinates of the cyclic curve are reduced 21 times in Fig. 52, 42 times in Fig. 53, 63 times in Fig. 54, 84 times in Fig. 55; and are thus not directly comparable with one another or with the companion curves.

"Returning to the spark-effect and the gradient curves,

however, it is seen that the spark-effect is by no means a simple function of the gradient of the cyclic curve. The curves show, moreover, that the maximum effect of the spark is greater in the cycles of larger amplitude; and not only because the hysteresis curves are steeper in large cycles than in small, but also because in small cycles the effect of the oscillations is, for some unknown reason, much smaller than the diminished gradient would lead one to expect.

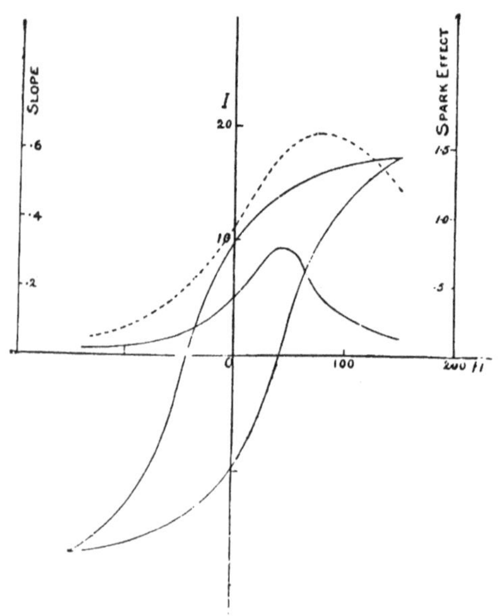

FIG. 54. The dotted curve shows spark-effect.

"It must be mentioned here that during the above measurements the effect of changing the sign of the initial charge given to the oscillation-coil was repeatedly tried. On no occasion was there any perceptible difference in the magnetometer deflexion. The damping of the oscillations must therefore have been very trifling."

Sella-Tieri Detector.—An interesting type of detector

has been invented by Sella, and improved by Tieri. In the latter form it consists of a core of soft iron wires, soldered together at their ends, and placed in a glass tube rather shorter than the core. On the outside of the tube are wound coils which are connected to the aerial and earth, and to a telephone respectively. A current is passed longitudinally through the core, while at the same time it is continually

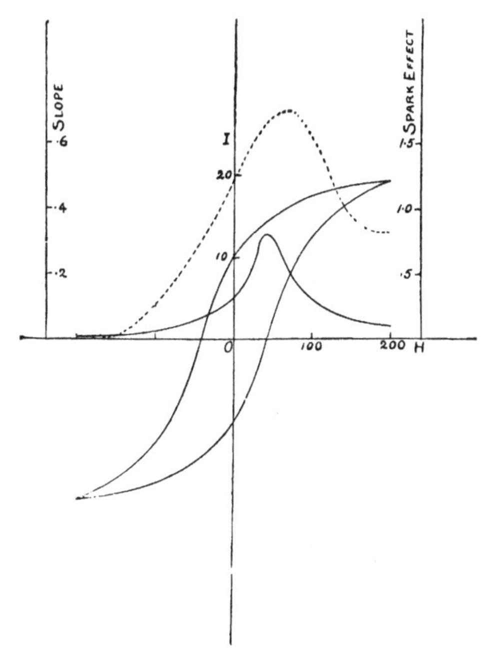

FIG. 55. The dotted curve shows spark-effect.

twisted and untwisted by a mechanical arrangement. The core is thus circularly magnetised by the current, and under the continually varying torsion is sensitive to the oscillating longitudinal field produced by the received oscillations. The telephone indicates the sudden alterations in magnetisation which occur on the reception of a signal from the transmitter.

CHAPTER VI.

THERMOMETRIC DETECTORS OF OSCILLATORY CURRENTS OF HIGH FREQUENCY.

THERE are a number of instruments in use which measure the energy of the oscillatory currents in the receiving circuit, and not the voltage or current as do coherers or magnetic detectors. Their action depends on the rate at which heat is produced in a short fine wire through which the oscillatory currents of high frequency pass. (I am tired of writing this long phrase, and its equivalent for dielectric currents, viz., "a train of electric waves of high frequency." The word oscillation is used by some writers to imply high frequency oscillation only. This is unfortunate, since even *electrical* oscillations have been produced at frequencies of less than one per second. I therefore propose to adopt a good old English word which signifies a periodic motion of high frequency, to stand for "a damped train of electrical oscillations of a frequency of the same order as is employed in wireless telegraphy;" or the corresponding "oscillatory currents, voltages, or magnetic fields of high frequency," associated with them in the sending or receiving circuits of a wireless installation. The word chosen is brief and descriptive, viz., **jig**. It may be qualified where necessary to prevent confusion, thus—an *electric jig*, a *magnetic jig*, an *Irish jig*—all implying periodic motions of very high frequency. A transformer of the high frequency currents used in wireless telegraphy has been called a jigger, hence also it seems natural to call these currents "jigs.")

The energy of electric jigs, then, may be measured by

their heating effect on a conductor. In some types of thermometric detector the temperature variation of the resistance of the fine wire in which the energy is converted into heat, is used as indicator; in others the heat from the wire acts on a thermo-electric junction which forms part of a galvanometer circuit. A third class, suitable for currents of some magnitude, utilises the expansion of the conductor. To the first class belong Professor Fessenden's original barreter, and Lieutenant Tissot's bolometer bridge; to the second Mr Duddell's thermo-galvanometer, and many others; and to the third that class of thermo-ammeters, whose construction is based either upon the direct elongation of a wire, as in the Cardew voltmeter, or on the untwisting of a spiral spring like the Ayrton and Perry instruments. There is also a fourth class, much used in Germany, in which the heating of the wire is measured by expansion of air in a vessel enclosing the wire.

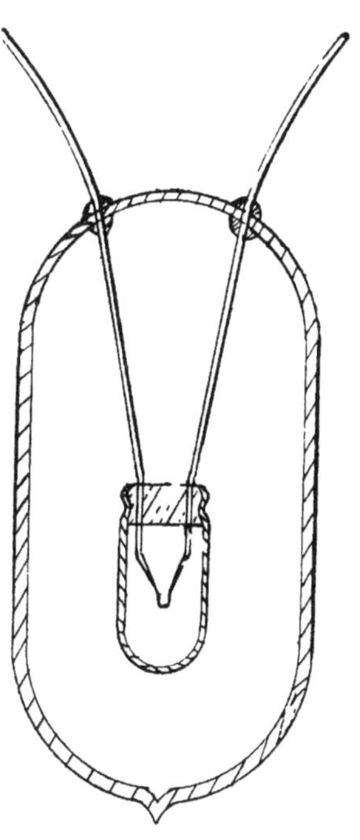

FIG. 56.—FESSENDEN'S BARRETER OR THERMAL DETECTOR.

The thermometric method is, so far, the only one which lends itself to accurate measurement of the current or energy, and as the rate of heat production is proportional to the square of the current, the thermometric detector can be easily graduated to show the R.M.S. current directly. Coherers and vacuum tubes act when the voltage rises sufficiently, magnetic detectors when the current attains a certain value, but the thermo-galvanometer indicates throughout its whole range the actual R.M.S. value of the current. In 1902 Professor Fessenden patented a form of thermal receiver which proved to be of the same order of sensibility as other wireless detectors. A short piece of extremely fine platinum wire, enclosed in a vacuum, is connected to the aerial and earth, and also forms one side of a type of Wheatstone bridge. The current jig in the receiver heats the fine wire, and as this has very small heat capacity and radiating surface, and is *in vacuo*, its temperature rises rapidly. The rise of temperature causes an increase of resistance which is indicated on the Wheatstone bridge by means of a syphon recorder or otherwise.

The system has been found to be very reliable in action. but from the fact that it depends on the energy received and not simply on the voltage or current, it is not likely to be able to compete with other forms of detector over great distances, for if the R.M.S. voltage of the jigs received falls off in proportion to the distance, the energy of the current jigs induced in the receiver will decrease in proportion to the square of the distance. As a detector, therefore, its efficiency will vary inversely as the square of the distance, instead of inversely as the distance. I understand that Professor Fessenden has now discarded this detector, at least for long distance transmission, in favour of an electrolytic arrangement which will be described in the next chapter.

Tissot's Bolometer Bridge.—Lieutenant C. Tissot, of the French Navy, has devised a kind of bolometer for use in the measurement of the received current in wireless telegraphy, and has made many useful measurements by its means. He has kindly allowed me to quote the following from his paper communicated to the Institution of Electrical Engineers in the beginning of 1906: *—

"The principle of Langley's bolometer is well known. Two fine metal wires are inserted respectively in the two

* By kind permission of the Council.

arms of a Wheatstone bridge. A variation in temperature of one of the wires produces a variation in its resistance which is indicated by the bridge galvanometer, the bridge having been previously balanced.

"In applying the bolometer to the detection of electric waves, it is necessary to ensure the complete heat isolation of the bolometric arms, and on the other hand to localise the action of the waves in one arm only.

"To effect this the arms, which are straight and very short—1.5 cm. of 10 μ diameter platinum wire, in the most sensitive models—are brought very near to one another within the same enclosure.

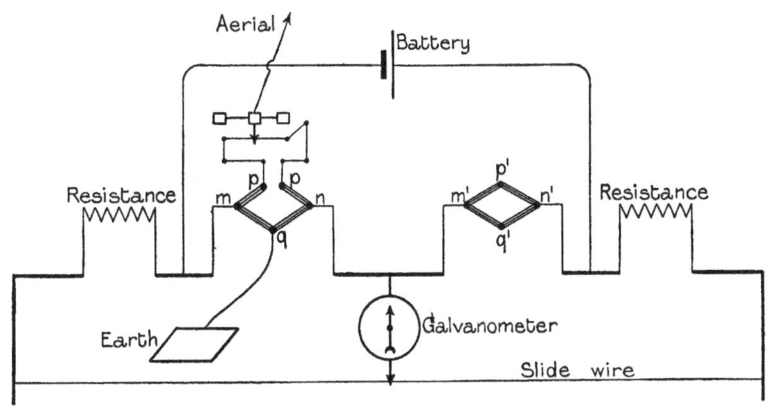

FIG. 57.

"In one of the types employed, these arms are *in vacuo*. The case in which they are contained is made as small as possible, and is enclosed in two successive coverings of silver-plated brass, between which is a very thin air space. The whole is immersed in a small vessel filled with water. In another model the heat isolation is obtained in a more simple manner by means of a Dewar vacuum vessel.

"According to the kind of measurements for which the apparatus is intended, two different methods are employed for localising the effect of the wave. In one of these methods, similar to that employed by Rubens, each arm of the bolometer is formed by four exactly equal pieces of wire arranged in the form of a bridge (Figs. 57 and 58).

"The balancing resistances of the bridge are either of German silver or platinoid, and are immersed in petroleum. The balancing of the main bridge is performed by means of a slide wire of large diameter. The aerial and the earth connection are attached at p, p, and q to that diagonal which is not in the circuit of the main bridge.

"The apparatus can be calibrated direct by a continuous current, and can be used as a wattmeter (the resistance being known and the self-induction negligible). The

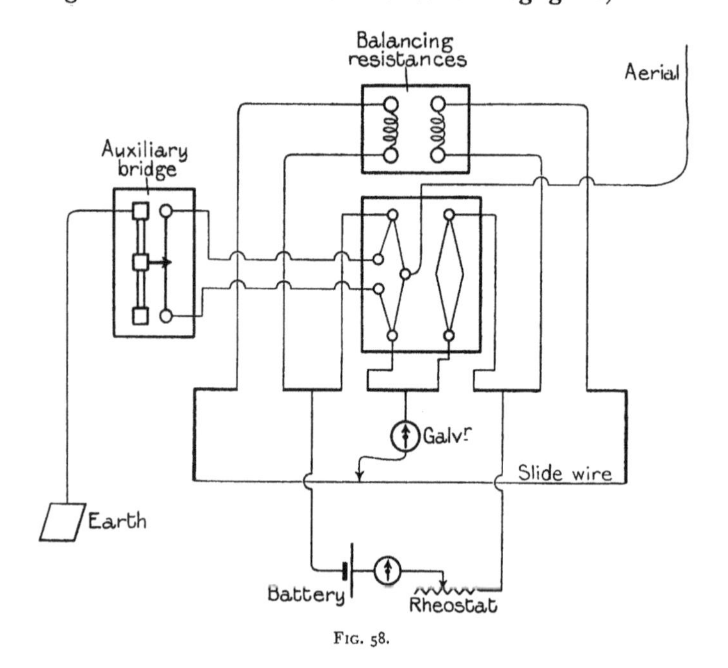

FIG. 58.

method employed for taking the measurements consists in the observation of the permanent deflection of the galvanometer of the main bridge under the action of the waves received during a suitable time. The aerial and the earth are removed, and there is connected in their place a source of direct current capable of supplying the required current to the bolometric bridge $m\,n\,p\,q$, so as to produce the same deflection of the galvanometer of the main bridge.

"Since it is necessary that the unbalancing of the main

bridge shall be solely due to the heat developed in the auxiliary bridge *m n p q*, an external means of adjustment (a kind of slide wire) was added to the bolometric bridge in order to be able to realise exactly the desired conditions. These are obtained when the galvanometer deflects in the same direction and exactly to the same amount, on reversing the direct current in the auxiliary bridge.

"The other method of localising the effect of the waves consists in inserting between the bolometric arms, each of which are formed of a single piece of wire, suitable ironless choking coils, of dimensions previously determined by experiment. The aerial and earth are then connected as shown in Fig. 59. The sensibility of these arrangements

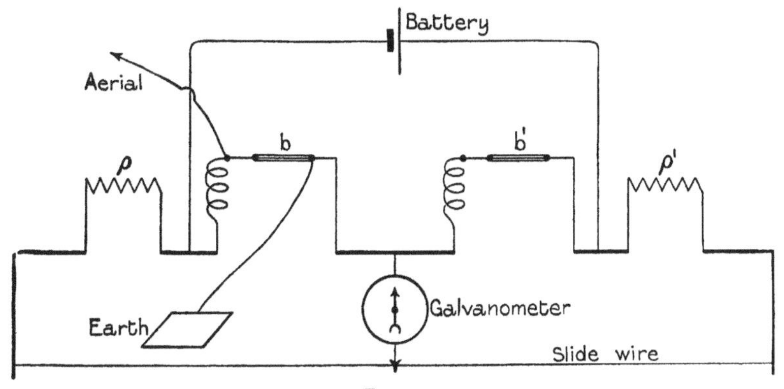

FIG. 59.

depends, of course, upon that of the galvanometer employed. For moderate degrees of sensibility a D'Arsonval dead-beat galvanometer was used. The apparatus is then suitable for use on board ship. Where extreme sensitiveness was required I employed a moving needle galvanometer (Thomson's type) with two parallel vertical needles, the resistance of the galvanometer being chosen equal to that of each of the bridge arms. I was thus able to obtain deflections of about 10 millimetres with an effective current of 100 microamperes (scale at 1 metre distance).

"If the length of the transmitting aerial A is left constant, and that of the receiving aerial B be varied progressively, it will be noted that the deflections of the bolometer reach a maximum for a certain length of B. The aerial system AB are then in resonance and have the same natural period.

"When the aerials A and B have the same shape, for instance both being either simple wires, or both consisting of four parallel wires, it will be found that the resonance always occurs when the lengths are equal whatever may be the general curvature or inclination of the aerials. The measurements given refer to systems in resonance."

Professor Braun of Strasburg and other German workers have made considerable use in experimental work of a thermo-ammeter of the air-thermometer type, designed originally by Mr Snow Harris. Like all other thermal instruments, it measures the steady temperature attained by the wire when heat production and radiation have reached a balance. A fine short wire is enclosed in air in the sealed end of a U-tube; a liquid is put in the lower part of the U, and the other branch is graduated. The heat expands the air, and the rise of the liquid column indicates the temperature, and therefore the current. This instrument is simple in construction, but several corrections require to be applied to the readings before the actual value of the current can be obtained.

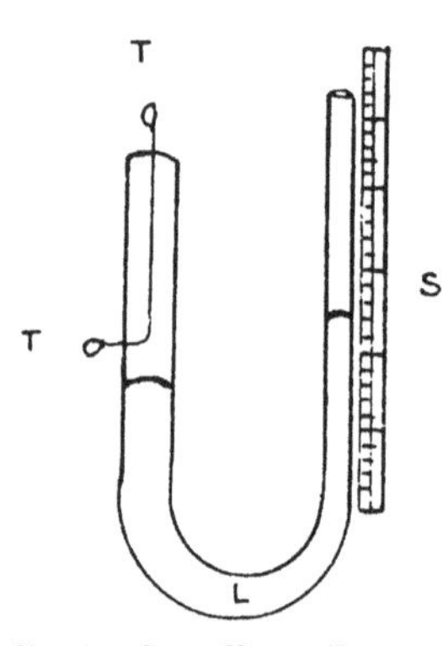

FIG. 60.—SNOW HARRIS THERMO-GALVANOMETER.

T, T, Terminals of fine wire; *L*, Liquid in U-tube; *S*, Scale.

Duddell's Thermo-galvanometer and Ammeter.—In order to measure the current given by a high frequency alternator, giving about 120,000 ~ per second, Mr Duddell devised a pair of instruments which between them could register through an enormous range, from about a micro-ampere (one millionth of an ampere) up to many amperes.

They are portable and easily adjusted, and are, so far, the best instruments to be had for the measurement of high frequency currents or voltages.

With Mr Duddell's kind permission,* I take the following descriptions from his paper read before the Physical Society and published in the *Philosophical Magazine*, July 1904:—

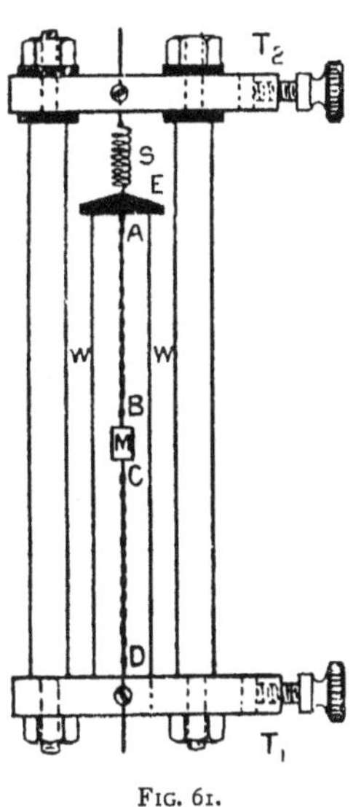

Fig. 61.

"The first instrument was designed to be very quick in action for a thermal instrument, and was made to work with a small current so that it could be used as a quick-acting voltmeter. It is essentially a very delicate Ayrton-Perry twisted strip-ammeter which has been improved by the addition of a temperature-compensation device to minimise the zero-creep when the temperature of the whole instrument changes. The working parts of the instrument are shown in Fig. 61. A B C D is the Ayrton-Perry twisted strip, of which the part A B is twisted in one direction and the part C D in the opposite. A mirror M and a very thin mica damping-vane are fixed at its centre. This strip is stretched in a frame formed of a brass block T_1 carrying one terminal and a piece of ebonite E, the sides of the frame being formed of wires W W. This frame itself is stretched by means of a spiral spring S fixed to the other terminal block T_2.

"The twisted strip A B C D and the wires W W are made of the same metal; in fact, the strip A B C D is actually flatted from the wire W W, so that the twisted strip and wires have the same temperature-coefficient of expansion. If the wires and strip rise in temperature equally, then the whole frame E W T_1 W simply gets longer and no twist of the mirror ensues. If, however, a current be sent from T_1

* By kind consent of the publishers.

to T_2 through the strip, then it heats and twists up, rotating the mirror M. Owing to the fineness of the strip (0.001″ Pt Ag) which is heated by the current, the instrument is very quick in action. The mechanical periodic time is also very short, about $\frac{1}{15}$ second; so that it is able to follow with a fair accuracy currents which vary over a small range as rapidly as one or two cycles per second.

"The data of this actual instrument are:—Resistance, 20 ohms; current to give 25 cm. at 100 cm. scale-distance, 22×10^{-3} ampere; P.D. to give 25 cm. at 100 cm. scale-distance, 0.44 volt.

"So that taking 1 cm. as the smallest measurable deflexion and 0.1 mm. as the smallest detectable movement—

The smallest measurable current is 4.4×10^{-3} ampere.
" detectable " 0.5×10^{-3} ampere.
Watts required to produce smallest measurable deflexion, *i.e.*, 1 cm., 387 micro-watts.

"It is evident that this very simple instrument has many uses. It is very easily set up, requiring no careful levelling, and is quite robust. The present instrument which I made three years ago has often been carried about just as it is in the pocket. The self-induction of the wire is extremely small, and the temperature-coefficient of the resistance of the wire (″ PtAg) is small also; so that the instrument can be used as a voltmeter to measure voltages down to about 0.1 volt, and I have used it in series with high resistances to measure voltages up to 10,000 volts. The instrument can of course be shunted to measure large currents; but if a good deflexion of say 25 cm. is required, the drop in volts on the shunt, 0.44 volt, is serious. Its main defect is that it requires screening from quick vibrations of the order of $\frac{1}{15}$ second, as the damping is not quite sufficient. I find that this can be easily done by standing the instrument on a heavy block which is suspended by means of wires and springs as in the Julius suspension for galvanometers.

"The second instrument is much more delicate and sensitive. It was made primarily to measure very small voltages and currents even with very high frequencies up to 120,000 ~ per second. At this high frequency the alternator I made would only give an extremely small output; it was therefore necessary, in order to carry out the experiment, to reduce the power required to work the instrument to its lowest possible value. I have called this instrument a Thermo-galvanometer.

"The principle of the thermo-galvanometer is quite simple. It consists of a resistance which is heated by the passage of the current to be measured, and the radiant heat from which falls on the thermojunction of a Boys radio-micrometer.* As at first constructed, it consisted of a heating resistance made of three or four turns of 0.001 inch diameter platinum-silver wire wound on a piece of mica, and placed as close as possible to the radiation receiving-plate of an ordinary Boys radio-micrometer, made by the Cambridge Scientific Instrument Company. This instrument was sensitive but exceedingly slow in action, taking over a minute to approximately attain the deflexion corresponding with the current flowing.

"A new radio-micrometer was therefore constructed following the instructions given in Professor Boys' paper, but having a very much smaller suspended loop than the radio-micrometer as usually constructed. The radiation receiving-plate was also omitted in order to reduce the quantity of metal to be heated, and the heating resistance was placed directly underneath the lower thermojunction, so that the junction received heat from it both by radiation and convection. The arrangement of the instrument is shown diagrammatically in Fig. 62.

"In the field between the pole-pieces N S (Fig. 62) of a permanent magnet is suspended by means of a quartz fibre a single-turn coil or loop of wire *l*, to the lower ends of which is fixed the thermocouple Bi, Sb. This loop is surmounted by a glass stem *g* and a mirror M. Below the lower junction of the thermocouple is fixed the heating resistance *h*, one end of which was connected to the frame of the instrument to avoid electrostatic forces. The action of the instrument is as follows. The current to be measured flows through *h* and raises its temperature, causing the lower junction of the thermocouple to rise in temperature above the upper, thus producing a current round the loop, which is deflected by the magnetic field against the torsion of the quartz fibre.

---

* *Philosophical Transactions of R.S.*, 1889, vol. clxxx. p. 159.

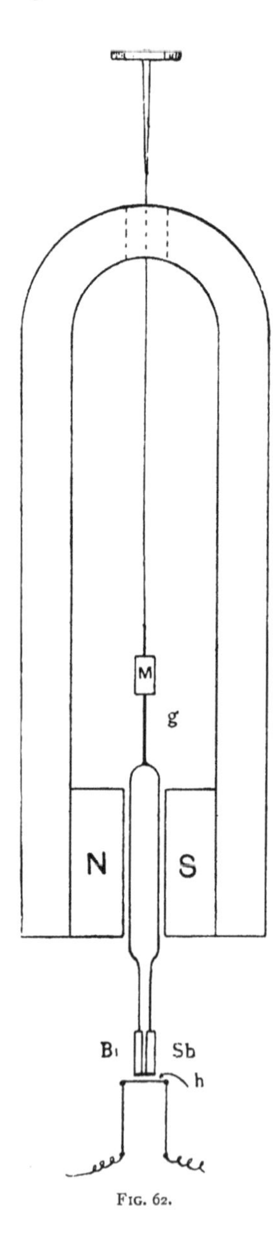

Fig. 62.

"The data of the instrument are:—Resistance, 18 ohms; current to give 25 cm. at 100 cm. scale distance, 8×10^{-4} ampere; P.D. to give 25 cm. at 100 cm. scale-distance, 14.4×10^{-3} volt.

"So that, taking as before 1 cm. as the smallest measurable and 0.1 mm. as the smallest detectable deflexion, the smallest measurable current is 1.6×10^{-4} ampere; the smallest detectable current is 0.2×10^{-4} ampere; watts required to produce smallest measurable deflexion, 0.46 microwatt.

"The deflexions of the instrument were nearly proportional to the square of the current; assuming the instrument correct at 100 divisions deflexion, then higher deflexions required to be increased by amounts gradually increasing up to between 4 and 5 divisions at a deflexion of 400 divisions, in order that the corrected deflexion might be accurately proportional to the square of the current flowing.

"To give an idea of the time the instrument requires to take up the permanent deflexion corresponding with any current, Table I. is given below of the deflexions at different times after starting a current of 8×10^{-4} ampere; and after interrupting the same, to show the time required for the instrument to return to zero. From which it will be seen that the instrument attains after 10 seconds its full deflexion to within 1 part in 500, or as it is a square law instrument indicates the true current to within 0.1 per cent.

"TABLE I.

Time in Seconds after Starting a Current of $8 \times 10^{-}$ Ampere.	Deflexion in Scale-divisions.	Time in Seconds after Interrupting the Current.	Deflexion in Scale-divisions.
5	500.5	5	1.5
10	501.0	10	1.0
30	501.5	30	0.5
60	502.0	60	0
120	502.0	120	0
180	502.0	180	0

"TABLE II.

For deflexion of 25 cm. at 100 cm. scale-distance.

Instrument.	Resistance in Ohms.	Current in Micro-amperes.	P.D. in Milli-volts.	Power in Micro-watts.
Thermo-galvanometer, gold heater	18	800	14.4	11.5
Thermo-galvanometer, platinum on glass heater -	103	346	35.6	12.3
Do. do.	202.5	275	55.6	15.3
Do. do.	363	231	84	19.4
Do. do.	1,071	121	130	15.7
Do. do.	3,367	88	296	26.0
Do. do.	13,910	31	431	13.9
Ayrton-Perry twisted strip - -	20	22,000	440	9,680

"This instrument, which forms practically a sensitive alternate-current galvanometer, worked very satisfactorily, even with the highest frequencies used of 120,000 ~ per second. As examples of its sensibility, it may be mentioned that on making a suitable noise into a Bell *telephone-receiver* sufficient current is generated to send the spot off the scale; and that if the thermo-galvanometer be connected to the line wires of a *microphone-transmitter* arranged in the ordinary way, whistling at a distance of from 15 to 20 feet from the microphone will cause deflections of the instrument of several hundred scale-divisions.

"A set of high-resistance heaters have been made and put in the above thermo-galvanometer, and the results obtained are given in Table II. The currents, P.D.'s, and power are those required to give a deflection ¼ the scale-distance, which forms a very convenient basis on which to compare square law instruments. I have also included the Ayrton-Perry twisted strip instrument in the table for comparison."

The last class of thermal detectors or ammeters, which depend on the expansion of the heated wire itself, are so simple in construction and so like instruments that are already well known to electricians, that it is unnecessary to describe them here. They are not sufficiently sensitive to serve the purpose of detectors in long-distance transmission. The Duddell thermo-galvanometer, on the other hand, though somewhat slow in action, is comparable in sensibility with the coherer, and can be used in telegraphy over considerable distances. As the readings of a thermal instrument are proportional to the squares of the currents, it is clear that it is much more sensitive in the upper part of its scale than in the lower.

On the whole, hot-wire detectors are not much used for actual telegraphic work, but almost all the measurements of the electrical energy concerned in the different processes of wireless telegraphy have been made by their aid.

CHAPTER VII.

ELECTROLYTIC DETECTORS.

ELECTROLYTIC detectors may be divided into two chief classes, both of which are in very common use in wireless telegraphy. The first class is a type of anti-coherer; *i.e.*, the resistance is increased by the action of the jigs; the second class is a species of intermittent electric valve, whose action resembles very closely that of the Wehnelt interrupter. Theory lags far behind practice in these matters, so the first thing to be done is obviously to give an unexplained description of the actual instruments. Of the first class the detector of Dr Lee de Forest is one of the best and most sensitive.

It consists of a tube containing two metallic electrodes, with an electrolysable paste of litharge with glycerine or vaseline mixed with water or alcohol, and a few metal filings. On the application of a continuous E.M.F., from a local battery, in series with a resistance of 30,000 or 40,000 ohms, a small continuous current passes through the paste, forming in it the fine crystalline structures usually called "lead trees." The aerial and earth wires are also attached to the terminals of the detector. On arrival of an electric jig the resistance momentarily increases, and the sudden change of current resulting is observed as a click in a telephone receiver in the local circuit.

From microscopic observations, Dr de Forest discovered that the treelike crystals, which are formed by electrolysis from various metallic solutions, are broken up by the action of current jigs, but are immediately reformed by the current from the local battery. It seems probable that the tips of the lead trees are melted or deflagrated by the oscillatory currents. The leaves of the trees are extremely thin plates, coming to very sharp points; as therefore the resistance must be almost entirely localised at the tips of the leaves, a very moderate amount of current will be sufficient to fuse them and make the circuit discontinuous. The rapidity with which the circuit recovers continuity proves that the breaks in it must be very minute, the crystals growing again immediately under the influence of the constant voltage applied by the local battery. This detector has been found to be exceedingly sensitive and quite reliable.

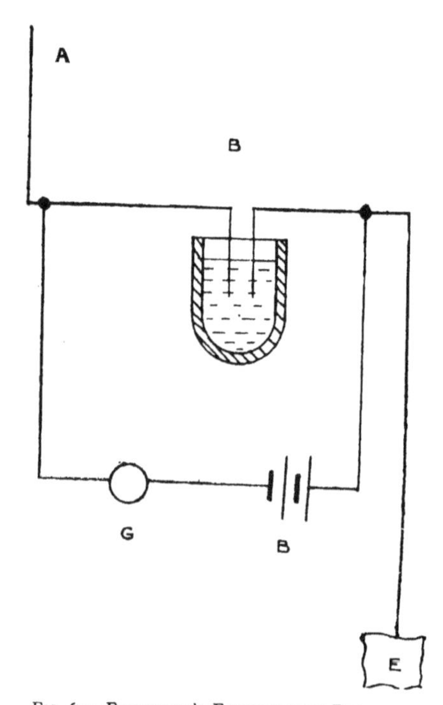

FIG. 63.—FESSENDEN'S ELECTROLYTIC RECEIVER, OR BARRETER.

A, Aerial; *B*, Barreter; *E*, Earth; *B*, Battery; *G*, Galvanometer or Telephone.

The second principal type of electrolytic detector has only a liquid between the electrodes, one at least of which must be very small. The best electrolyte for signalling purposes appears to be nitric acid. A battery and a telephone receiver are put in series with the detector. If the terminals of the detector are also connected to air and earth wires, signals may be received as with other forms of detector. The invention appears to have been made independently by Fessenden, Ferrie, Schloemilch, and Vreeland.

The thing is practically a small Wehnelt interrupter, in which the applied voltage of the battery is too small to maintain an appreciable current without the assistance of the electric jigs. A Wehnelt break will not start work unless there is a certain minimum inductance in circuit. This condition means that the break will not start unless the electricity is surging to and fro, since the effect of putting inductance in a circuit is to make such revibration possible. Under these conditions quiet electrolysis of the liquid gives place to a much larger but intermittent current, suitable for use in the primary of an induction coil. Perhaps the theory which best suits both cases is, that the surges destroy the "polarisation" of the cell, and thus allow a sudden rush of current from the battery which shortly fails again owing to reformation of the film which constitutes "polarisation." The sudden stoppage produces again electrical surges in the circuit if the total resistance be low, as in the Wehnelt break, which again destroy the polarisation. In the detector, or barreter, as Professor Fessenden calls it, the applied voltage is not high enough to keep up the intermittent current by causing surges in the circuit when it stops and starts, the current therefore only flows when surges are impressed upon it by an outside agency, *i.e.*, the jigs transmitted to the detector by the aerial and earth wires.

There has been considerable difference of opinion as to the mode of action of these detectors, Professor Fessenden holding that it is due to variation of the resistance of the electrolyte through heating by the current, while others speak of "polarisation" as the only cause. One difficulty in deciding between the claims of the rival theories is the vagueness of the knowledge which we have of the actual mechanism of polarisation; also it must be observed that a thin film of vapour produced by heat, and not electrolysis, would appear to be equally effective in increasing the resistance, and may account for the action of the Wehnelt break. It seems possible that the actual molecular process in the barreter may be otherwise explained by taking into account

Dr H. S. Sand's observations on the rapid increase of resistance which takes place at an electrode very shortly after a direct current is started in an electrolyte. Dr Sand * has shown that even in a cell in which polarisation does not, under ordinary conditions, occur, the resistance of the layer of electrolyte nearest to the electrode is much less at the commencement than it is when a steady state has been attained, and he has proved that this rise of resistance is due to the layer next the electrode becoming partially

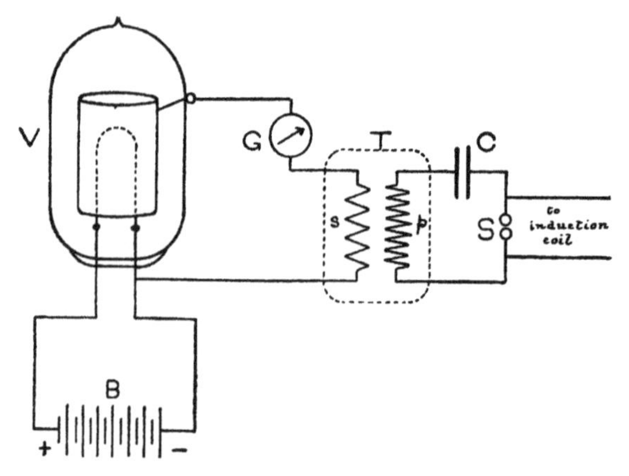

FIG. 64.—VACUUM VALVE USED TO RECTIFY ELECTRIC OSCILLATIONS AND RENDER THEM DETECTABLE BY AN ORDINARY GALVANOMETER, *G*.

(From "Electric Wave Telegraphy," by kind permission of Prof. Fleming and Messrs Longmans, Green, & Co.)

depleted of the ions which were in it before the E.M.F. was applied. In the barreter this action must be superposed on any others which are taking place, and may be the most important of them all. The jig voltage, by creating ions may cause increased conductivity, which is immediately lost by continued application of the steady voltage of the local circuit. It is thus possible that the phenomenon may be explainable without reference to films of gas or even temperature variations of resistance.

* *Proc. Faraday Soc.*, vol. i. p. 1, 1904.

This type of detector is very sensitive, and is in use in many wireless telegraph stations in all parts of the world.

Fleming's Hot Carbon Rectifier.—Professor Fleming discovered in 1890 that a current will pass with ease from the glowing filament of an incandescent lamp to another conductor in the bulb. If the second conductor be a piece of metal the current will pass from the metal to the carbon through the space only when the positive terminal is connected to the metal and the negative terminal of the cell to the negative end of the battery which heats the carbon filament. The arrangement is thus a rectifier, *i.e.*, if an alternating voltage be applied to the terminals, an intermittent current in one direction only will be transmitted. The apparatus may be used to detect oscillatory currents by means of an ordinary galvanometer, as under favourable circum-

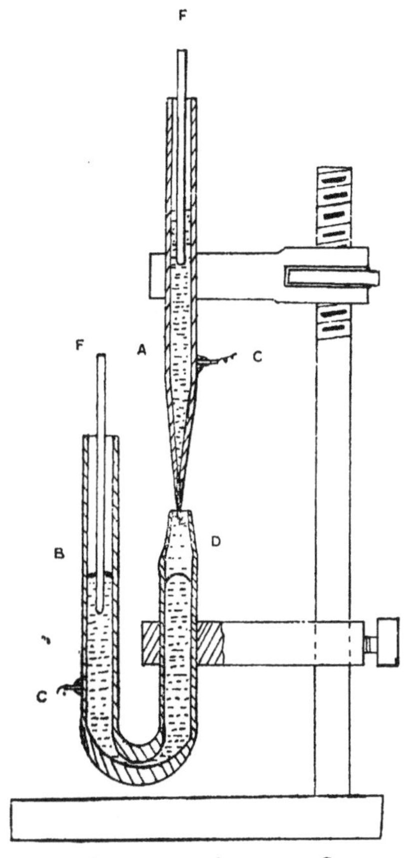

FIG. 65.—ORLING AND ARMSTRONG CAPILLARY RECEIVER:—ADAPTED FORM OF LIPPMANN'S ELECTROMETER.

A, Glass tube ending in fine point, filled with mercury; *B*, U-tube with mercury:—dilute acid at *D*; *C*, *C*, Platinum wires connecting mercury with aerial and earth; *F*, *F*, rods for regulating mercury level.

stances, *e.g.*, when the carbon filament is giving light at the rate of about one candle per three watts and the metal conductor is cool, the rectification is stated to be from 80 to 85 per cent.

Armstrong and Orling Electro-capillary Detectors. —Messrs Armstrong and Orling have shown that a capillary electrometer of Lippmann's invention may be used as a detector of jigs. A glass tube, drawn down to a very fine bore at one end, dips into a wider tube. Mercury is put in the wide tube with dilute sulphuric acid on top of it. Mercury is also put in the upper and wider end of the fine tube, but does not run out at the bottom because of its surface tension. The lower end of this tube dips into the sulphuric acid in the wide tube, and by forcing the mercury down the upper tube and allowing it to retract, the air in the upper tube is driven out and the acid electrolyte becomes continuous between the mercury electrodes. Very minute changes of voltage will cause an alteration of the surface tension of the mercury, and produce a motion of the column in the fine tube, which is easily observable by means of a magnifying glass, and may be recorded photographically or otherwise.

I have included the above descriptions in order to show by what a large variety of means it is possible to detect electric jigs. Success in wireless telegraphy, then, depends much more on the proper adaptation of the detector, of whatever kind it may be, to the rest of the apparatus, than on the actual species of detector used.

CHAPTER VIII.

THE MARCONI SYSTEM.

WITHOUT going into questions of priority of invention, we shall commence our description with Marconi's system, as it was probably the first open-circuit telegraph in commercial use. Already, in Chapter II., the general principle of the original invention has been described; we must now go into details of the more modern apparatus.

The more important improvements embodied in recent installations are (1) the receiving transformer, or jigger; (2) the adoption of Tesla's method of producing long trains of oscillations of high frequency and voltage; and (3) the magnetic detector. Other improvements consist in methods for the employment of an alternator of considerable size in place of the modest induction coil, and in various combinations of inductances and capacities for the purpose of rendering the receiver responsive to jigs of a definite frequency only. Similar additions have been made to the apparatus employed in almost every other system. The questions of priority of invention have not in most cases been settled by the courts as yet. Many difficulties have also occurred on account of the enormous power of the spark in a large station, for though the total amount of energy sent out at each spark may be comparatively small, perhaps only a few foot-pounds, the time is so short that the activity during the existence of the spark is very great. In a small station which sends out about one foot-pound of energy per spark, the horse-power during the spark is about 300; in a larger station it is proportionately greater, probably amounting to tens, or even hundreds, of thousands of horse-power. This high rate of working, though only momentary, requires that

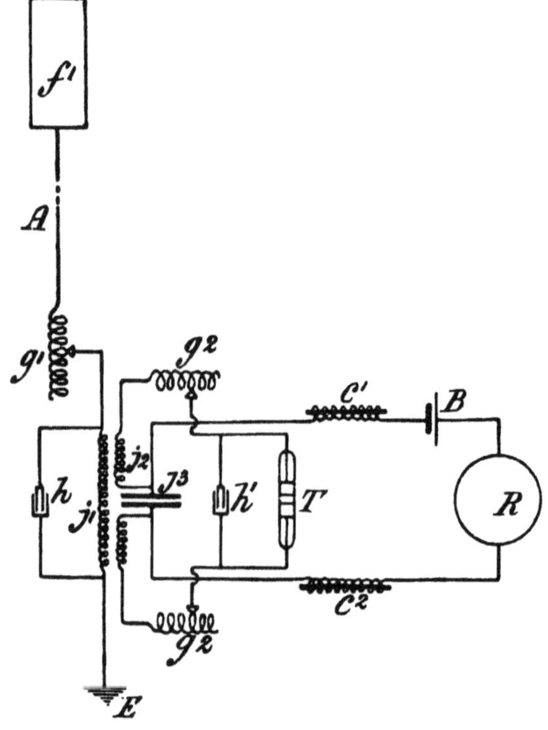

Fig. 66.—Arrangement of Receiving Apparatus in Marconi System of Syntonic Wireless Telegraphy.

A, Antenna; *E*, Earth-Plate; g^1, g^2, Tuning Inductance; j^1, j^2, Jigger; j^3, Jigger Condenser; c^1, c^2, Choking Coils; *T*, Sensitive Tube, or Coherer; *R*, Relay; *B*, Battery.

(Note.—Figs. 66 and 67 are inserted by kind permission of Prof. Fleming and Messrs Longmans, Green, & Co.)

conductors and insulators alike should be of special designs best suited to the new conditions. The determination of the proper forms for the multifarious apparatus necessary has been a cause of delay in the completion of some of the larger stations to an extent which was not foreseen. The preliminary difficulties have, however, been got over satisfactorily, and for some time past the transmission of messages over distances of at least two thousand miles has been carried on with regularity. Fig. 66 shows diagrammatically the arrangement of the receiver in a modern long-distance station on the Marconi system.

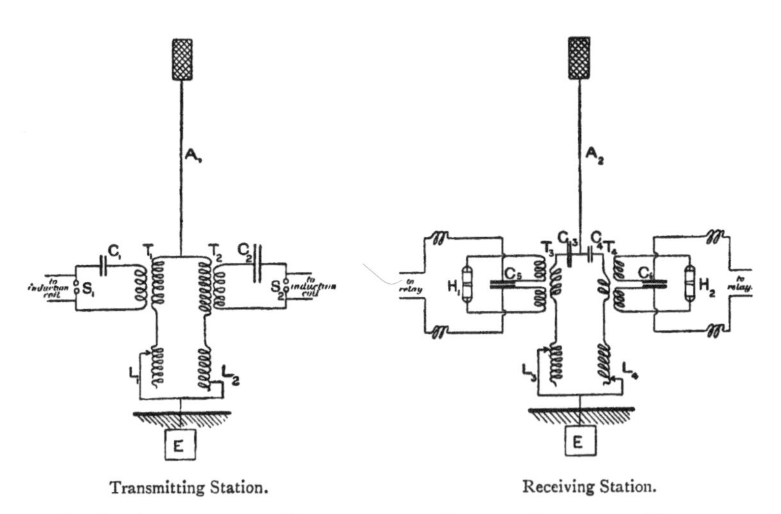

FIG. 67.—ARRANGEMENT OF TRANSMITTING AND RECEIVING APPARATUS IN MARCONI SYSTEM OF MULTIPLE SYNTONIC WIRELESS TELEGRAPHY.

Among other details of the apparatus an important one is the sending key patented by the Marconi Company, in conjunction with Mr Andrew Gray. In dealing with the large alternating currents in use at the larger stations it was clearly impossible to use an ordinary Morse key for transmitting on account of the arc which may be formed between the contacts. To obviate this difficulty the key shown in Fig. 68 was designed. It is constructed so as to allow of the contact being broken only at a moment when the instantaneous value of the current is zero. As this occurs perhaps a hundred times every second the actual break does not lag appreciably behind the movement of the handle by the operator. The contacts are held together by an electro-magnet actuated by the alternating current, and therefore are only free to separate at the moment when the value of the current is zero, which happens twice during every alternation.

The spark itself was at first the cause of much trouble in high power stations, the great heat generated producing pits in the metal of the discharging balls at the point where a spark took place. The deafening rattle of the discharge was also very trying to the operators. To obviate both

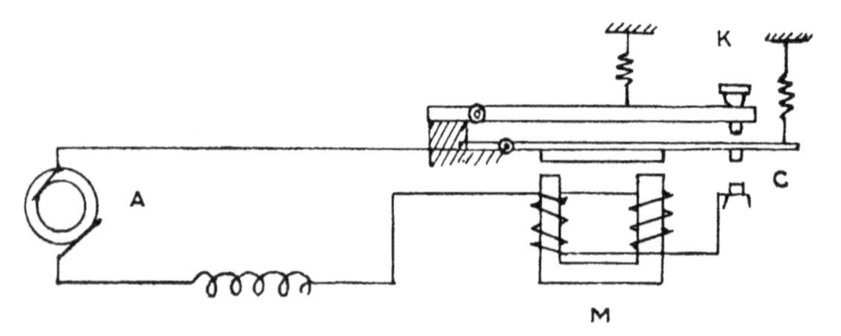

FIG. 68.—GRAY'S SENDING KEY FOR ALTERNATOR CIRCUIT (Marconi Company Patent).

A, Alternator; *M*, Electro-magnet; *C*, Circuit-making Contacts; *K*, Handle of Key. The contact *C* is only broken by the action of the spring at the moment when the magnetic attraction of *M*, and therefore the main current, is zero, thus avoiding arcing.

difficulties Professor Fleming designed a cast-iron box in which the spark balls were placed, being mounted so that they could be kept in rotation by a motor outside the box. The box was filled with compressed air to increase the efficiency of the spark in producing oscillations. The principle of the apparatus, except in regard to the revolving balls, is the same as that of Mr Fessenden's earlier American patent.

The outside gear at a large station is shown in Fig. 69. The pyramidal aerial depicted, though of large capacity and convenient shape, has the disadvantage that the waves from one side interfere to some extent with those from the opposite side, since their phases on arriving at the receiver are slightly different. This was pointed out by Dr de Forest, who found that a fan-shaped aerial radiates much

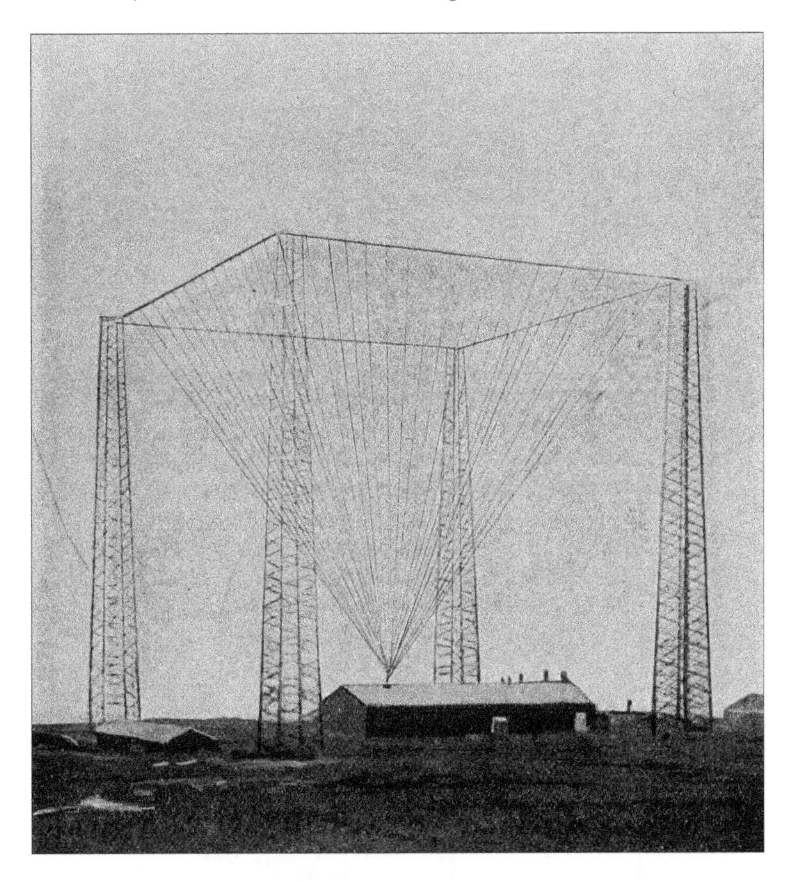

FIG. 69.—MARCONI STATION AT GLACE BAY.

more strongly in a line perpendicular to its plane than in the plane itself, and that the same is true as regards absorption of jigs by a fan-shaped aerial at a receiving station. We shall return to this matter in Chapter XV.

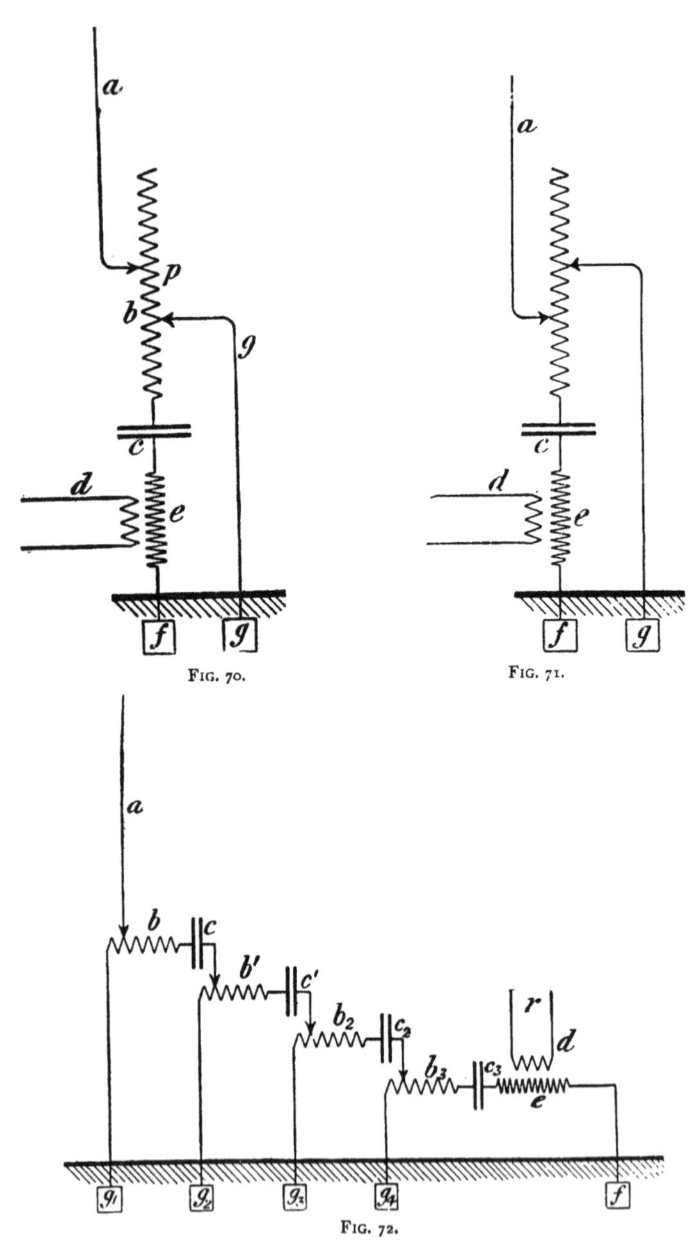

FIG. 70.

FIG. 71.

FIG. 72.

DIAGRAM SHOWING THE CONNECTIONS IN MARCONI'S X-STOPPER.
a, Receiving Antenna; *p*, Inductance Coil with Sliding Contact, *b*; *c*, Condenser; *e*, *d*, Jigger in Cymoscope; *f*, *g*, Earth-Plates.

Another improvement in the Marconi system is the X-stopper, an instrument which is intended to transmit only jigs of a definite frequency from the aerial and earth terminals of the receiver to the detector. This is accomplished by making two earth connections to the aerial. One of these, which contains much capacity and inductance, and has therefore a definite frequency of vibration, is the receiver circuit; the other earth wire has very little capacity or inductance, and will therefore conduct jigs of any period as forced vibrations. If jigs having the proper frequency are received on the earth and aerial wires, they cause the receiver circuit to revibrate, and are recorded in the instruments; if disturbances of any other frequency arrive they do not go through the receiver circuit, but pass between the earth and aerial by the second wire whose electrical inertia is so much smaller.

The best results are obtainable by using a number of the capacity and inductance vibrators in series, each with its proper earth wire, the receiver being on the last vibrator, as in Figs. 70-72.

The X-stopper (which means the stopper of stray waves or X's) should be compared with Slaby's aerial and syntonic horizontal wire with two earth-plates, which it resembles. Other arrangements of capacities and inductances for a similar purpose have been patented by Professor Fessenden and almost every inventor in wireless telegraphy. The magnetic detector and other improvements now embodied in Mr Marconi's system have been already described.*

---

* For further details of the Marconi system the reader is referred to "The Principles of Electric Wave Telegraphy," by Prof. J. A. Fleming (Longmans).

CHAPTER IX.

THE LODGE-MUIRHEAD SYSTEM.

SIR OLIVER LODGE, realising more fully than other inventors the importance of having the transmitter and receiver of the same natural frequency, patented in 1897 a system of syntonic wireless telegraphy. The apparatus was based directly on Hertz's radiation experiments, and on his own resonating circuits and point coherer.

In Chapter II. a sketch has been given of the earlier forms. I shall now describe very fully the system now in use commercially in many parts of the world, availing myself (through the courtesy of the Lodge - Muirhead Syndicate and the proprietors of the *Electrician*) of the subjoined description by Mr H. C. Marillier, which appeared (in substance) in the *Electrician* of 27th March 1903.

Mr Marillier commences his paper with a brief reference to the preliminary tests to which the system of wireless telegraphy devised by Sir Oliver Lodge and Dr Alexander Muirhead had been subjected, with the result (he says) that the inventors now feel justified in offering it as a real advance upon any of the systems hitherto put forward, as regards clearness and accuracy of signalling.

"This claim," he goes on to say, "is well borne out by the accompanying specimen of script (Fig. 73), taken under ordinary working conditions from a siphon recorder coupled directly to a new pattern of coherer without any relay and without any tapping back apparatus, a simplification which in itself constitutes a marked and definite improvement.

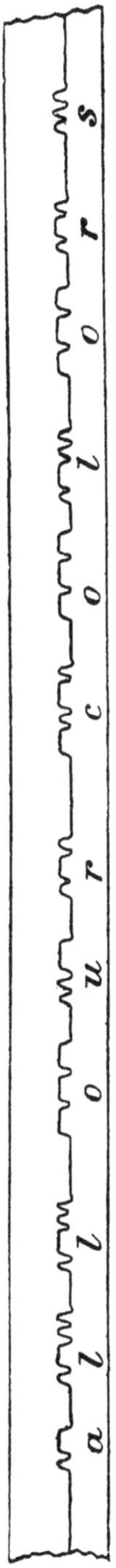

Fig. 73.—Alternator, 600 frequency, about 15 Words per Minute.

"The fundamental patents on which the Lodge-Muirhead system depends were taken out for the most part in 1897, and provide for the following essentials:—

"1. The combination, in the transmitting and receiving circuits, of two capacity areas and an inductance coil, as a vital element in a syntonic system of wireless telegraphy. The capacity areas may be regarded as the two coats of a Leyden jar spread out in space, one of them being suspended in the air and the other near the earth, in fact, the earth itself if desired. Between them is situated the spark-gap, and between the spark-gap and the lower capacity area is the inductance coil as well as a condenser, both being adjustable (Fig. 74).* The purpose of these is to prolong the electrical oscillations, and by means of their adjustment to tune the radiator to any desired frequency or pitch, and thus render syntony in the receiver possible. Sir Oliver Lodge's main idea in regard to transmission has always been to obtain a succession of true waves of definite frequency, the cumulative influence of which will produce a perceptible effect on a suitably tuned receiver, no matter how feeble the waves may be, after the well-known principle of sympathetic resonance. This action is illustrated electrically in Sir Oliver Lodge's old experiment of the syntonic Leyden jars, in which the closed circuit of one jar responds to the other, so that the jar overflows or gives other signs of having become charged whenever discharge occurs in the second through its own distinct, similarly tuned circuit. The adjustability of the inductance and con-

* In Figs. 74-77, and 83-85, the top connection a leads to the upper, and that from x or x_1 to the lower insulated capacity area of the oscillator.

denser render it further possible to tune the two circuits in such a way that they shall be protected against certain kinds of specified outside interference; but for very close tuning of this kind more elaborate devices are necessary, which will not be entered into in the present article.

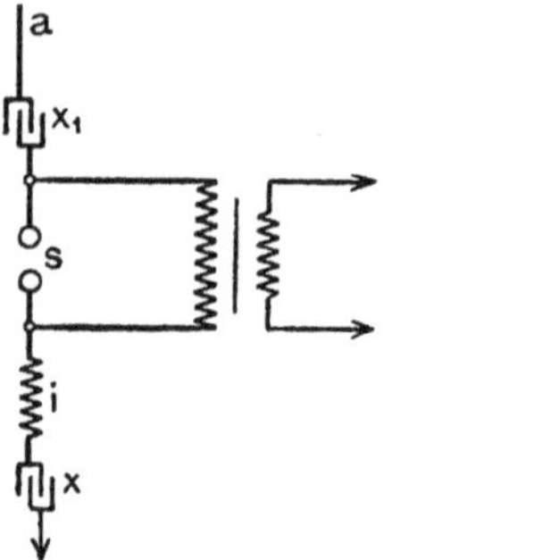

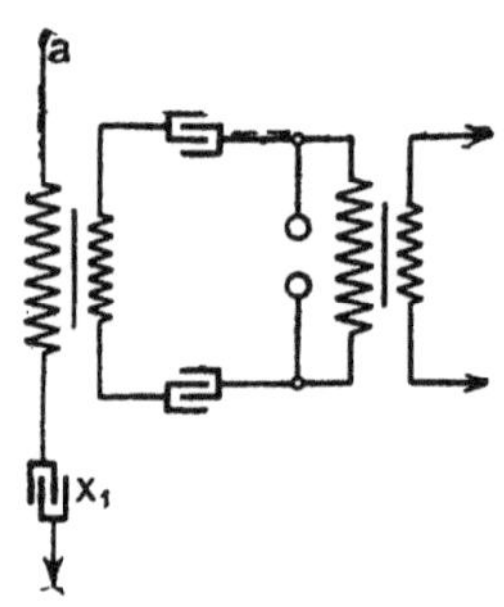

FIG. 74.—OPEN SENDING CIRCUIT. FIG. 75.—CLOSED SENDING CIRCUIT.

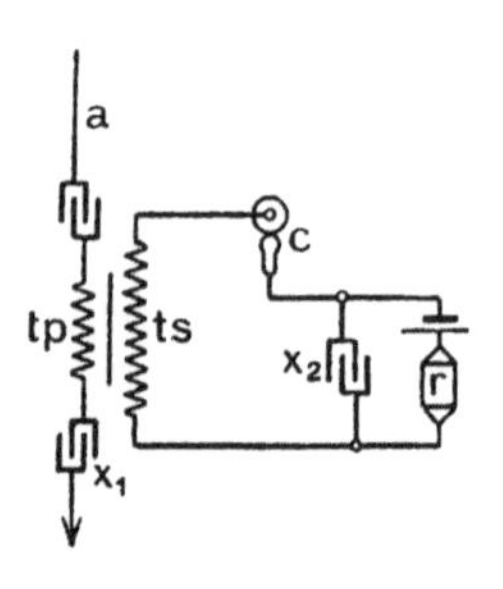

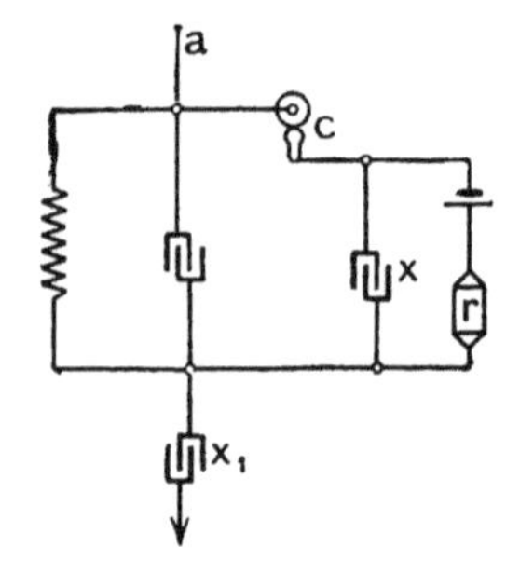

FIG. 76.—OPEN RECEIVING CIRCUIT. FIG. 77.—OPEN RECEIVING CIRCUIT.

"2. The second point claimed by the inventors is the use of a transformer, or ironless induction coil, in the receiving circuit (see *tp*, *ts*, Fig. 76). This device is probably essential in all long-distance wireless telegraphy, and should disputes arise at any future time as to the claims of individual inventors, it is a feature which might conceivably be fought over. Its purpose may be described briefly as tending to magnify the E.M.F., and putting the coherer into a second circuit, instead of in direct series with the vertical collecting wire and the lower capacity. The passage in which this transformer claim occurs is as follows, and may be found in Sir Oliver Lodge's patent, No. 11,575 of 1897:—

"'In some cases I may . . . surround the syntonising coil of the resonator with another or secondary coil (constituting a sort of transformer), and makes this latter coil part of the coherer circuit, so that it shall be secondarily affected by the alternating currents excited in the conductor of the resonator, and thus the coherer be stimulated by the currents in this secondary coil, rather than primarily by the currents in the syntonising coil itself; the idea being thus to leave the resonator freer to vibrate electrically without disturbance from attached wires.'

"The illustration which accompanies this description in the specification definitely indicates the transforming-up and magnification of the E.M.F.

"3. The use of a condenser shunt (x_2 in Fig. 76) in the coherer circuit, such as will enable that circuit to have a definite time period, as described in the Lodge-Muirhead patent, No. 18,644 of 1897. The addition of the condenser as a shunt eliminates the battery and receiving instrument so far as oscillations are concerned, and is regarded by the inventors as a feature of great importance. This, as well as the transformer, has been adopted in principle by other workers, and a common modification of it is to divide the secondary of the transformer in two parts and place the condenser between them. In the Lodge-Muirhead patent, No. 29,069 of 1897, the condenser is shown placed between the lower or earth capacity and the syntonising inductance coil described in 1. In practice, when a transformer is used, it has been found best to have two condensers, the one described, and another in the secondary transformer circuit (see Fig. 76).

"The last-mentioned patent also provides for enclosing the coherer in a completely closed metal case, a precaution which, in the practical development of the system, is supplemented by a device which automatically short-circuits the coherer whenever the connections are switched from the receiving on to the sending apparatus.

"Besides these more or less essential features of the system, Sir Oliver Lodge and Dr Muirhead have introduced a number of detailed changes into the apparatus which tend to simplify or to improve the working. The most important of these is the new coherer. Up to a short time ago the inventors thought that they had reached the desired point of perfection in a filings-tube coherer fitted with two needle points in place of the usual plugs or knots, the tube being tapped by a constantly working clockwork tapper as shown by Sir Oliver Lodge in 1894. There are certain disadvantages, however, attendant upon all coherers in which metal filings are employed. They require a special and carefully adjusted apparatus for tapping back, and many of them are capricious in working and liable to fatigue. The new coherer introduced by Lodge and Muirhead requires no tapper, but is kept perpetually sensitive by the rotation of a small steel disc just separated from a column of mercury by a film of mineral oil. The impulse of electric oscillations breaks down the oil film and establishes momentary cohesion between the steel disc and the mercury. The coherer, without its metal case, is shown in Fig. 78, and its construction in Figs. 79 and 80, in which *a* is the rotating steel wheel, and *b* the mercury in its trough *d*, with an amalgamated spiral of platinum wire *c*, connecting it to the terminal or binding screw *h*. *e* is a copper brush making connection with the steel disc *a*, through its axle *j*. *f* is a spring carrying a small cushion of felt *k*, which rests lightly on the steel disc for the purpose of keeping its edge clean and free from dust before and after contact with the mercury. *g* are ebonite wheels which gear the steel disc to the clockwork movement actuating the siphon recorder, to the side of which the metal case containing the coherer is fixed.

"The coherer, as already mentioned, is placed directly in circuit with a siphon recorder, without the interposition of any relay, and is connected also with a potentiometer for the purpose of regulating the P.D. at its terminals from 0.03 to 0.5 volt, according to circumstances. The coherer is so sensitive, when the wheel is rotating with moderate speed, that the application of a whole volt is sufficient to break down the film of oil and establish coherence. Hence only a fraction of a battery must be used, with the rest of the circuit of appropriate low resistance; and it can readily be understood how the slightest electrical surging, producing any electrostatic effect whatever, is sufficient instantly

FIG. 78.—COHERER.

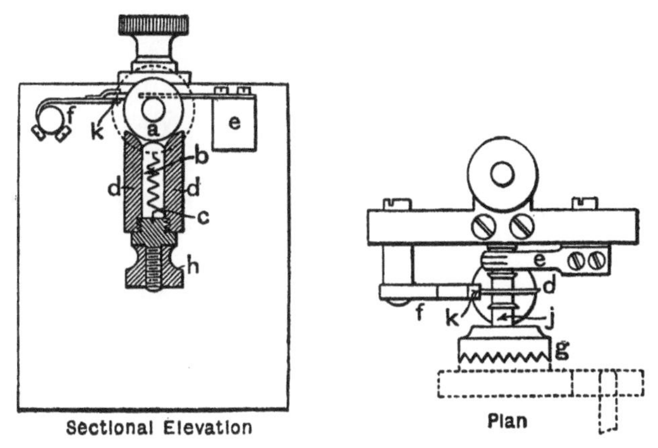

FIGS. 79, 80.—COHERER, IN ELEVATION AND PLAN.

FIG. 81.—THE "BUZZER."

to break down the film, start the connection, and give the required signal. It is difficult to suppose that any recording instrument operating electrostatically can be more sensitive and more trustworthy than this simple device.

"The transmitting circuit in an ordinary open system consists essentially of an elevated capacity, preferably a large globe or a roof, although in general practice a wire cage arrangement, offering as little resistance as possible to the wind, may be used (see Figs. 86 and 88 to 91). From this capacity hangs the vertical wire, which is brought into the operating room and connected to one knob of the spark-gap. The inductance coil and condenser are connected in series with the other knob, and a lead is then taken outside again to the second capacity. The system also includes a Ruhmkorff coil for giving the sparks, and a battery of five cells, more or less, which is replaced

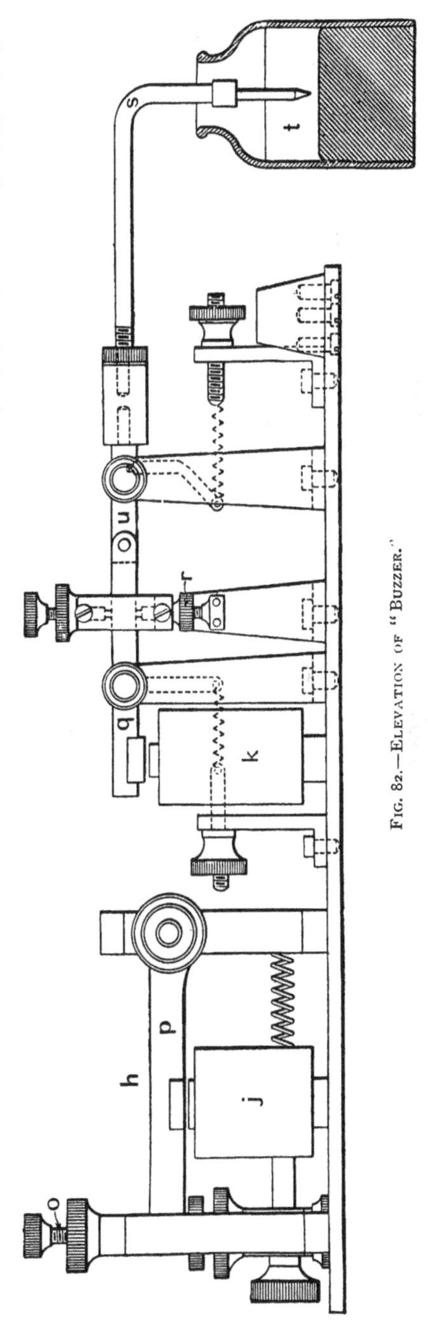

FIG. 82.—ELEVATION OF "BUZZER."

for long-distance work by an alternator. Various open and closed methods of making the connections are shown in Figs. 74, 75, 83, and 84.*

"The signalling apparatus consists either of a specialised

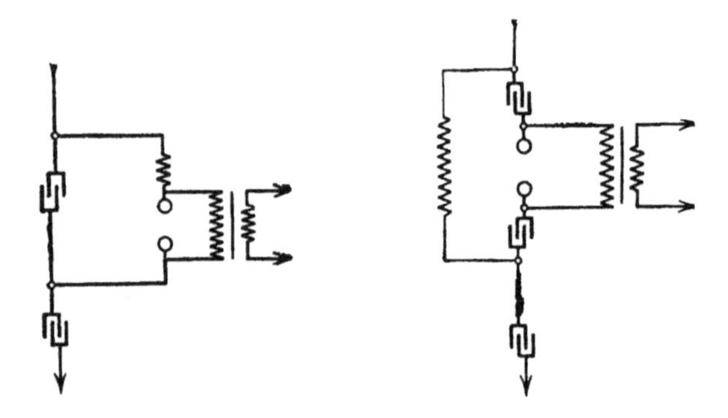

FIG. 83.—ANOTHER FORM OF CLOSED SENDING CIRCUIT.

FIG. 84.—ANOTHER FORM OF CLOSED SENDING CIRCUIT.

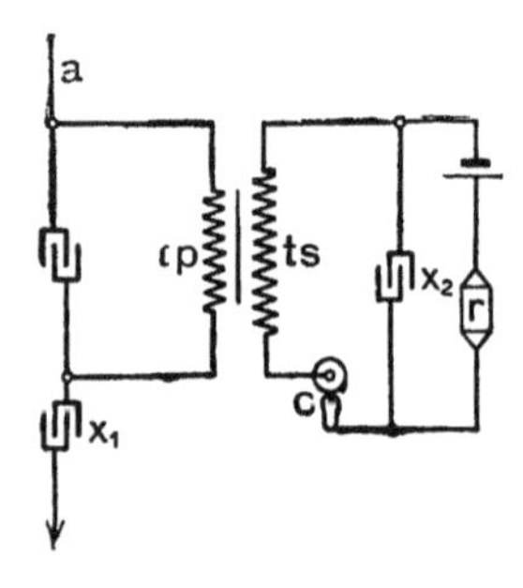

FIG. 85.—CLOSED RECEIVING CIRCUIT WITH TRANSFORMER.

Morse key, worked by hand, or, what is a distinctive feature of Dr Muirhead's improvements, an automatic signalling machine used in conjunction with a perforator of special pattern. In either case the local signalling circuit, as

* See foot-note on page 149.

arranged by Dr Muirhead, contains a transmitter designed to open and close at a definite rate the primary of the induction coil. This instrument, which is shown in Fig. 81, and more graphically in Fig. 82, is made up of two tele-

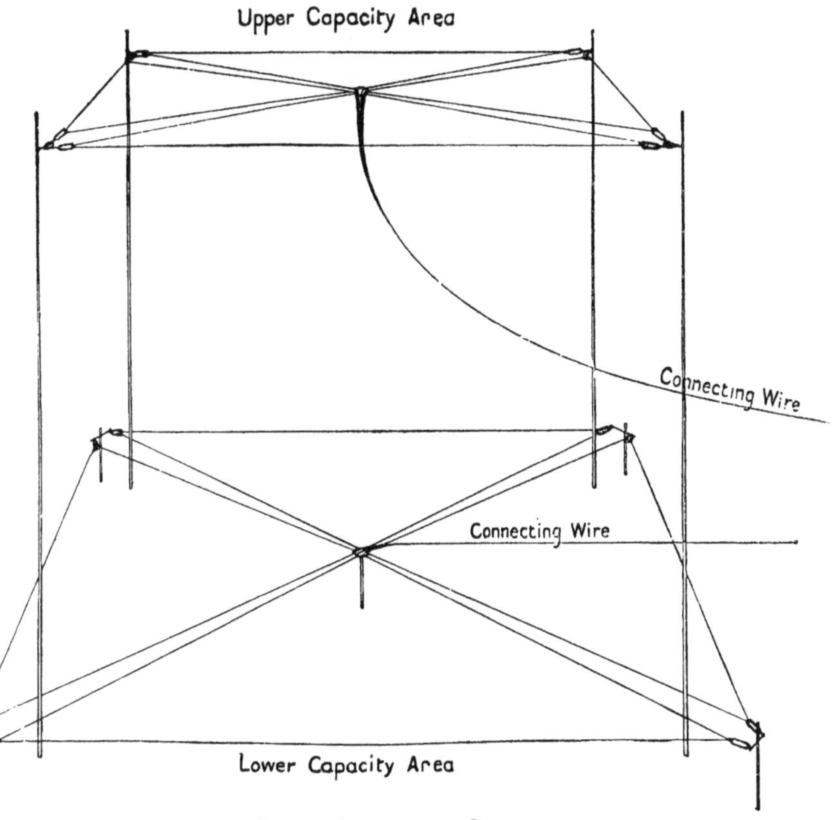

FIG. 86.—INSULATED CAPACITY AREAS FORMING OSCILLATOR AS USED IN 1907.
(Illustration kindly lent by the Lodge-Muirhead Syndicate.)

graphic sounders cross-connected in such a way as to act reciprocally. An aluminium arm fitted with a copper rod dipping into mercury is attached to the armature of the second sounder, and the rapid make and break between the copper rod and the mercury (about 600 times per

Fig. 87.—General Arrangement of Transmitting and Receiving Apparatus.

A, 12-volt Box of Secondary Cells; *B*, Complete Receiving Set; *C*, Spark Rod Frame; *D*, 10-in. Spark Coil; *E*, Buzzer; *F*, Perforator; *G*, Hand Sending Key; *H*, Primary Switch; *J*, Auto-transmitter.

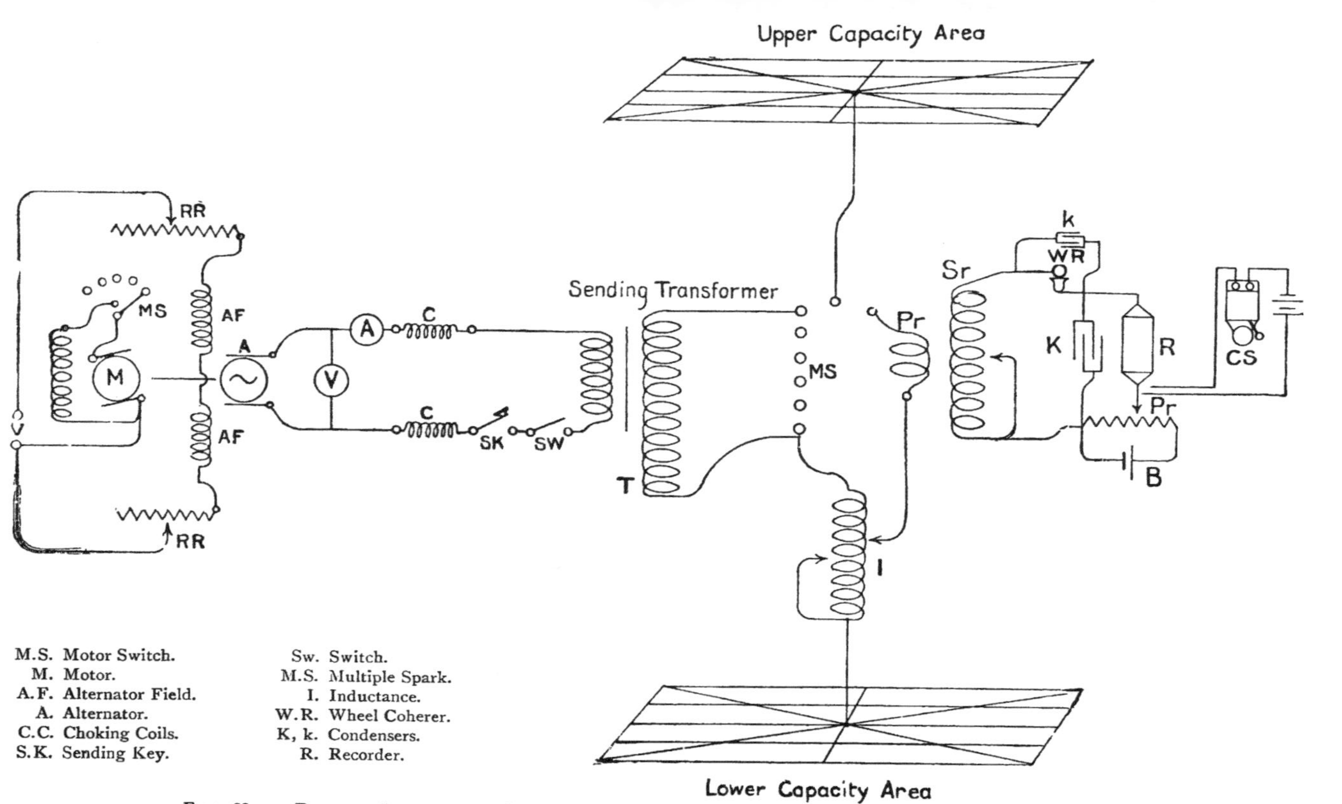

M.S. Motor Switch.
M. Motor.
A.F. Alternator Field.
A. Alternator.
C.C. Choking Coils.
S.K. Sending Key.

Sw. Switch.
M.S. Multiple Spark.
I. Inductance.
W.R. Wheel Coherer.
K, k. Condensers.
R. Recorder.

FIGS. 88-91.—DIAGRAM ILLUSTRATING COMPLETE TRANSMITTING AND RECEIVING STATION OF MOST RECENT TYPE.

minute) serves to fix the frequency of the sparks. The apparatus is known as a "buzzer," and its function is, during the holding down of the Morse key, to cut up the long-continued contact into a rapid succession of sparks, without any attention on the part of the operator (save in so far as he may first adjust the armatures of the sounders,

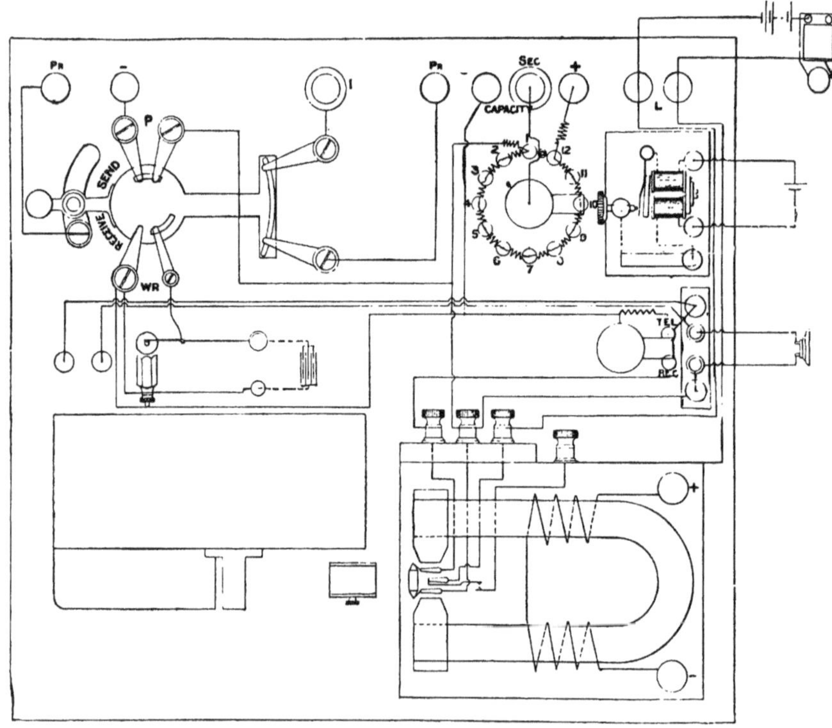

FIG. 92.—DETAILS OF RECEIVING APPARATUS.
(Compare with right half of Figs. 88-91.)

or the length of dip in the mercury, to suit conditions), so that all he has to do is to signal shorts and longs in the usual telegraphic manner. These are translated by the "buzzer" into the requisite mode of disturbance for spark-signalling, and are translated back by the recorder into signals of long and short duration on the tape (see specimen of script illustrated). If the rapidity of the sparks is insufficiently great, the signal becomes a broken or wavy line, and not a steady deflection; but if the rapidity be increased, the waviness disappears, and the recorder needle is simply held over, giving a steady long deflection to represent a dash, and a momentary one to represent a dot. This steadiness of deflection, though pleasing to the eye,

FIG. 93.—BACK VIEW OF RECEIVING APPARATUS.
K, Recorder; *L*, Coherer; *M*, Clockwork; *N*, Change-over Switch; *O*, Potentiometer *P*, Transformer.

is not actually essential to the reader who, with a little practice, can decipher the signals equally well when the sparks are so slow that three separate impulses or deflections represent a dash and one a dot. It may be observed, however, that the new coherer, in combination with a recorder, follows every fluctuation and peculiarity of the received waves, and indicates any inequality and uncertainty which may occur in the sparking transmitter. This method of reception is far preferable to one which acts only through the intervention of a contact-making relay, and is convenient for studying the changes which are brought about by every variation in the mode of signalling. If the signals are coming with irregular amplitudes, the legibility can be improved by limiting the excursion of the siphon recorder by the insertion of a stop, against which it shall strike, and be held over by a definite amount at each signal.

"The direct application of ordinary telegraphic signalling apparatus, both at the sending and receiving ends, is one of the most valuable features of the Lodge-Muirhead system, and is chiefly due to Dr Muirhead, who, indeed, foresaw its possibility as early as 1894, when Sir Oliver Lodge first demonstrated his early experiments before the Royal Institution."

Through the kindness of the proprietors of the *Electrician* I am enabled to make the following excerpts from the report of Mr M. G. Simpson, Electrician to the Indian Government Telegraph Department, on the establishment of wireless communication between Burma and the Andaman Islands. The report (dated 21st December 1905) forms an interesting record of a successful installation under all the adverse conditions of a tropical climate, and is a testimony to the practical value of the Lodge-Muirhead system:—

"Towards the end of August 1904 operations commenced for linking up the Settlement of Port Blair in the Andaman Islands with the general telegraph system of India by establishing wireless telegraph stations at Port Blair and at Diamond Island, which is already connected by a short cable to the mainland. The distance to be spanned is 305.2 miles. Between Port Blair and Diamond Island is situated Table Island, on which is a lighthouse and close to which ships pass on the voyages Rangoon and Colombo, and Calcutta to the Straits. It was therefore decided to take the opportunity of establishing a station on Table Island which it was hoped would prove useful to the shipping and to the Meteorological Department. This distance from Diamond Island to Table Island is 130.2 miles, and from Table Island to Port Blair

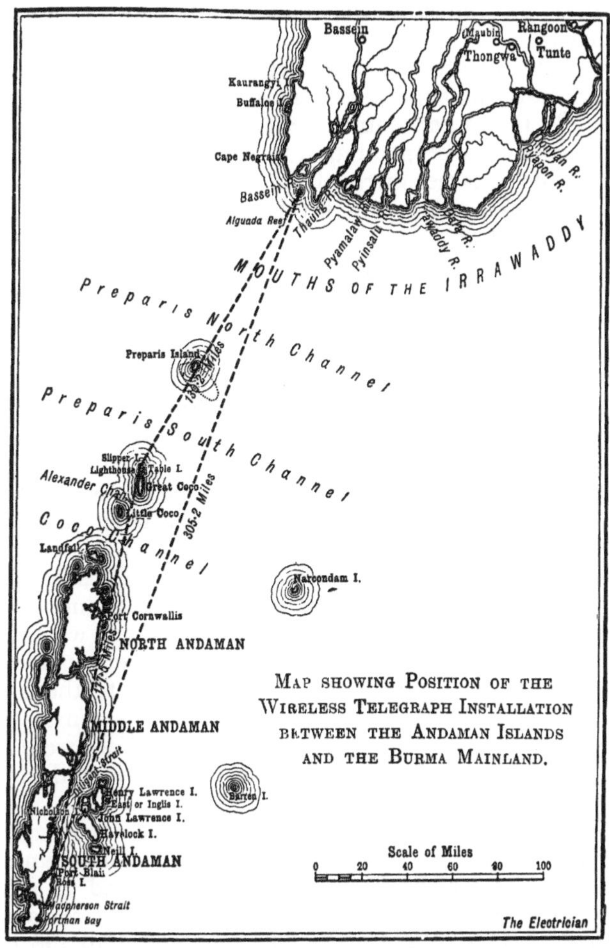

FIG. 94.—MAP SHOWING STATIONS FOR WIRELESS TELEGRAPHY BETWEEN ANDAMAN ISLANDS AND BURMA.

177.5 miles. A map is attached (Fig. 94) which shows the positions of the stations.

"The system chosen was the Lodge-Muirhead, with

which we had already experimented for three months in the early part of 1904. For this system four masts are required at each station, and as the distance to be spanned was over 300 miles, it was thought advisable to use 150 feet masts.

"I left Calcutta on 8th October 1904, with Mr J. N. Parker, assistant superintendent, one signaller, and native staff. By the same steamer went the three masts and all instruments and stores required for the station at Port Blair. The steamer arrived on 11th October, and the erection of the masts and fitting of the station was proceeded with at once.

"The Chief Commissioner placed at my disposal for the erection of the masts the services of Mr Bonig, assistant harbour master, with the requisite convict labour. The work was carried out by him with great expedition, and all four masts were erected complete by 28th October. The engine and dynamo were then fixed in position in their tent and the charging of the batteries started. The whole station was completed on 11th November and worked well with a temporary station erected on R.I.M.S. "Minto," which was lying up the harbour and was separated by a fair-sized hill.

"Leaving Mr Parker in charge of operations at Port Blair, I left on 7th November for Diamond Island, where I arrived on the 13th. The installation here was completed on 20th November, and attempts were made to communicate with Port Blair from the 20th to the 23rd, when a small motor broke down. As the wave measurer expected from England had not yet arrived, we were not in a position to tune the stations accurately, and could scarcely hope to attain success.

"I left Diamond Island on 25th November for Rangoon in order to take masts and stores to Table Island and to select the site for the installation. We arrived at Table Island on the morning of 30th November, and remained two days landing the stores. Having regard to the disposition of the neighbouring Cocos Island, it appeared desirable to place the wireless installation as much to the west as possible. With this object I decided to place the station, not on Table Island at all, but on Slipper Island, which is a quite small island just to the west of Table

Island, and connected to it at low water by a narrow strip of land. Having selected the site for the masts, I returned to Diamond Island, arriving on 2nd December, accompanied by Mr C. Landon, assistant superintendent, who had joined me in Rangoon.

"We now continued trials between Diamond Island and Port Blair, and though we got occasional signals, we got nothing readable till the early morning of 22nd December 1904, when for ten minutes we received the first readable signals from the Andaman Islands to Burma by wireless telegraphy. Considering that the distance between these stations is 305.2 miles, and the power used was 36 volts and about 15 amperes, or only a little over ½ H.P., this result is a remarkable proof of the efficiency of the Lodge-Muirhead system.

"Experiments on the direct circuit from Diamond Island to Port Blair could not at this time be continued, as Mr Parker left Port Blair on the 22nd to come to take charge of the station at Diamond Island while I went to Table Island to complete the installation on Slipper Island. Mr Landon accompanied me, and we arrived at Table Island on 28th December.

"There is no landing place for a boat on Slipper Island. There are two landing places on Table Island, one on the south side, to be used during the north-east monsoon, and one on the north side, to be used during the south-west monsoon. The landing is very difficult in the bad weather, and often in the monsoon is impossible. Our stores were all landed on Table Island, and had to be carried across to Slipper Island when the tide permitted. The masts for this station had already been erected, and the aerial was put up and all the instruments installed by 11th January. On 12th January messages were exchanged with Diamond Island, and communication established. This circuit continued to work well till the middle of May, when an important piece of machinery broke down which could not be repaired till towards the end of June.

"I left Table Island on 17th January, leaving Mr Landon in charge of the Slipper Island installation. I arrived at Diamond Island at 7 A.M. on the morning of the 18th. The wave measurer had at last arrived, and I had just time to measure the tune of the aerial on Diamond

Island before the steamer left at 10 A.M. for Port Blair, where we arrived on the 19th.

"On 30th January we were able to receive the signals both from Diamond Island, 305.2 miles, and from Slipper Island, 177.5 miles, at Port Blair, but those offices were at this time unable to receive the signals from Port Blair, though they were in communication with each other.

"It now only remained to put in at Diamond Island and Slipper Island the same tuning apparatus that we had installed at Port Blair. Mr Shields took the first available opportunity of proceeding to Diamond Island, where he arrived on 9th February, and on 10th February 1905 direct communication from Port Blair to Diamond Island was established, a number of messages being exchanged on that day. From this date on, communication had been kept up continuously, with the exception of certain interruptions to be mentioned hereafter, and the amount of traffic exchanged has been considerable, as will be seen from the statistics appended.

"At this time we were using for sending a 20-inch induction coil worked from an accumulator battery, but it was not expected this would prove powerful enough for continuous work, and 3 H.P. engines with alternators and transformers had been telegraphed for. For receiving we used telephones operated by carbon steel coherers. At first, on the circuit Slipper Island to Diamond Island, we used the Lodge-Muirhead coherer, operating a syphon recorder and giving signals on a tape, but our signallers, being always accustomed to read by sound, much preferred to read from the telephone, and the signals in a telephone operated by a carbon steel coherer were preferred by them to those in a telephone operated by the Lodge-Muirhead coherer.

"On 20th February the new 3 H.P. engines with alternators and transformers sent out by Messrs Muirhead & Co. arrived in Rangoon, and arrangements were made to get them out to the three wireless stations as soon as a steamer was available. They were landed at Diamond Island on 4th March, Slipper Island on the 5th, and Port Blair on the 6th. The engines and alternators were set up and running at Diamond Island on 6th March, and Port Blair on the 10th, and a marked improvement at once resulted in the signals, both on account of the increased loudness and improved tone.

"About this time occurred the first serious interruption. Diamond Island was unable for four days (10th March to 13th) to read the signals from Port Blair owing to atmospheric disturbances. During this time Port Blair was able to read Diamond Island signals quite well; but on a previous occasion (6th March) the reverse was the case, Diamond Island being able to read the signals from Port Blair, but Port Blair could not read the signals from Diamond Island. No serious trouble in working was experienced during the rest of March, but it was noticed that atmospheric disturbances came on nearly every afternoon, often making work difficult.

"Since April statistics were kept of the traffic exchanged; and a copy of the records for the months of April, May, and June is appended. The installation was only opened in a limited degree for State traffic, and the actual number of messages tendered and dealt with in April was only fifty-five sent from Port Blair, and thirty-nine received for Port Blair, but a great deal was sent in addition to these messages, and the traffic during the month amounted to nearly 10,000 words, and more could easily have been done had it been required. It will be noticed there were no interruptions during the month of April. During the month of May traffic exchanged amounted to nearly 14,000 words. The traffic during July, August, September, October, and November has amounted to 18,542, 21,286, 21,408, 27,398, and 26,938 words of actual messages. More could, of course, have been dealt with had it been tendered.

"It appears certain that atmospheric disturbances are more prevalent at Diamond Island than at Port Blair. The interruptions, however, to communication caused by them would have been greatly reduced, both in frequency and duration, if everything had been nice and quiet. Up to nearly the end of June the apparatus at Diamond Island was installed in a tent, and there is no doubt the signallers had to work under very great disadvantages. A perfect din was caused by the public works party erecting the permanent office building, which is fortunately completed now.

"Just as a land line is subject to occasional interruptions from falling trees, floods, and other physical causes, so a wireless installation of whatever system is subject to occasional interruption from electrical causes. And as in the former case experience has shown us how to so improve the methods of construction and maintenance of the land lines as to minimise the interruptions in number and duration, so it is to be expected that experience will show us how to improve our methods of wireless telegraphy to attain a similar result.

"The present installation we have must be regarded as being as fully reliable as an ordinary land line of about 300 miles erected through forest, and it could, I think, successfully challenge comparison with any other wireless installation of similar length in respect to regular continuous work.

"It may be of interest to note that the usual speed of working is from seventeen to twenty words per minute. In fact, it is limited only by the skill of the operator."

CHAPTER X.

THE FESSENDEN SYSTEM.

PROFESSOR FESSENDEN commenced in 1897 the development of the system which is now the property of the National Electric Signalling Company. For two years he was engaged by the United States Government for special research in the subject, and had the advantages of all the resources of a Government department at his command. His inventions are very numerous and, in many respects, original, and his results show a precision and practicality not attained by many other experimenters in the same field.

Magnetic, thermal, and electrolytic detectors, methods of exact tuning, and even wireless telephony are covered by Mr Fessenden's patents. Among these perhaps that which has contributed most to the success of the system is the barretter. In its original form this was a thermal receiver, depending for its action on the change of resistance in a very fine platinum wire when carrying the jig current. Latterly the continuous wire has been discarded in favour of an electrolytic cell, one electrode of which is an extremely fine point. The apparatus has been described more fully in Chapter VII. An important feature of this system which greatly aids secrecy of transmission is the arrangement of the sending key which does not break the circuit, but merely alters the wave-length of the waves given out by cutting out some inductance. Thus, unless a receiving station is tuned with extreme accuracy to the transmitter, it will receive instead of signals only a long

FIG. 95.—AERIAL CONDUCTOR (415 feet high), FESSENDEN SYSTEM, AT MACHRIHANISH.
A Steel Tube braced by steel wire guys divided by insulators into 50-feet sections. Capacity increased by network supported by branches near top of tube.

Fig. 96.—Fessenden's Machrihanish Station—Base of Aerial Conductor.
Showing insulating cup and ball joint supporting it, and sending and receiving leads.

continuous dash, hence only a very sharply tuned receiver will receive a message at all. In the latest forms of apparatus the difference in frequency between the waves sent out during spaces, and those sent as signals is only $\frac{1}{4}$ per cent.; interception by rivals is therefore almost impossible. Mr Fessenden is apparently the first to use an aerial consisting of a steel tube standing on an insulating foundation and held in position by insulated stays. Figs. 95, 96, and 97, which have been made from photographs kindly lent me by the National Electric Signalling Company, show the aerial and details of the network of earth wires, or "wave chute" at Machrihanish. When it is mentioned that the distance between successive insulators on each stay is 50 feet, the great dimensions of the tube will be more easily realised.

Many of the more recent improvements cannot be described in detail owing to the fact that the patents are not yet published. As, however, it is frequently of greater practical utility to have well-attested facts in regard to the performance of instruments whose details are unpublished, than to follow the more common plan of giving accurate descriptions of apparatus whose practical value is still a matter of conjecture, I append some notes of results, chiefly taken from reports by experts belonging to various branches of the United States Government service, which have been kindly given me by the representative of the National Electric Signalling Company in Scotland:—

"The National Electric Signalling Company's wireless apparatus comprises the following:—A. Sending and receiving apparatus; B. Interference preventer; C. Secrecy sender; D. Anti-atmospheric device; E. Intensity regulator; F. Wave measurer.

"To avoid misunderstanding, the capabilities of the various devices are set forth under the following headings:—

1. Best results which have been obtained.
   (*a*) In official tests by U.S. Army or Navy.

FIG. 97.—FESSENDEN'S MACHRIHANISH STATION.
Showing concrete anchor and strain insulators at end of stay, also network of earth wires (wave chute) on ground.

(*b*) In tests where the results have been thoroughly investigated and confirmed.

2. Results which may be expected in ordinary use.
3. Results guaranteed.

A. *Sending and Receiving Apparatus.*

"1. Hatteras Lightship No. 72, equipped with our apparatus, has carried on a telegraphic conversation with the U.S. Navy Station at Dry Tortugas, Fla., equipped with our liquid barretter. This distance is over 1,100 miles. Mast on lightship 75 feet high, power ½ K.W. Mast at Dry Tortugas about 180 feet high, power 3 K.W. Results obtained by Navy operators and reported officially. These results were only obtained during night time.

"U.S. vessels equipped with our apparatus lying off Norfolk communicate with more or less regularity with the stations around Boston. (Reports of Navy operators.)

"In the official tests by the U.S. Navy of the 1 K.W. sets installed on the 'Alabama' and 'Illinois,' communication was maintained over a distance of 268 nautical or 300 land miles, in daylight in the month of August, in spite of heavy atmospheric and other interference. Masts 135 feet high. Results obtained by Navy operators and embodied in official report.

"With land stations messages have been exchanged at night time between Machrihanish, Scotland, and Brant Rock, Massachusetts, a distance of 3,004 land miles.

"The results obtained with the Dry Dock 'Dewey' have been published by the Navy. The Hatteras Lightship was equipped with our apparatus complete, and the Dry Dock 'Dewey' with an infringing copy of this Company's liquid barretter and sending apparatus.

"2. With the standard 1 K.W. outfit, and masts 135 feet high, communication should be maintained under all conditions over a distance of 300 sea miles. With the 5 K.W. standard sets a distance of 500 miles should be maintained under all conditions, and with the 20 K.W. sets, suitable for installation on scouts, a distance of 800 to 1,000 miles.

"3. The distances guaranteed are as follows:—1 K.W. sets, 250 miles; 5 K.W. sets, 400 miles; 20 K.W. sets, 700 to 750 miles.

[The horse-power required therefore varies approximately as the square of the distance for distances up to 1,000 miles.—J. E.-M.]

B. *Interference Preventer.*

"1. Messages have been received with a loss of one word in twelve (easily corrected on repetition) with a 2 K.W. station operating only 226 yards away from the receiving station. (Official test.)

"Under the same conditions none of the other companies were able to receive when the interfering station was nearer than $5\frac{1}{2}$ miles. (Official report.)

"When the interfering ship was so close that its signals burned out filings of the coherer used by [another] company, and the sending ship was so far away that messages could not be received on the coherer, even when the interfering ship was not interfering, the installation of the interference preventer and liquid barretter enabled messages to be received with ease while the interfering ship was trying to interfere. (Official report.)

"In this case the interfering ship was about 500 yards away from the receiving ship, and the sending ship was about 120 miles away.

"During the combined manœuvres at Fortress Monroe, the U.S. Navy vessels attempted to interfere with the stations installed by this Company for the Coast Artillery. The interfering vessels were within 2,000 yards but were absolutely unable to interfere. (Official report.)

"While messages were being received at Brant Rock from Machrihanish, a 15 K.W. station belonging to a rival company situated at Boston thirty miles away, endeavoured to interfere by sending on a tune which was not more than 3 per cent. different from the tune at which Machrihanish was sending. The interfering station was cut out and messages received from Machrihanish.

"2. With old-type interference preventer, a difference in wave-length of 3 per cent. is as a rule sufficient to cut out interference, unless the interfering station is in close proximity, in which case a 10 per cent. difference in wave-length is necessary.

"By the use of this interference preventer the signals

are not weakened, but on the contrary, generally slightly increased in strength.

"With the latest improved type of interference preventer, a difference of ¼ per cent. in wave-length is sufficient for ordinary cases, and a difference of 3 per cent. where the interfering station is very close. In this case the strength of the signals is cut down to approximately ½.

"3. Guarantee. Where the distance of the interfering station from the receiving station is more than 1 per cent., of the distance between the sending and receiving stations. a difference in wave-length of 3 per cent. is sufficient to cut out the interference without the signals being appreciably weakened, interfering and sending stations being of equal power.

C. *Secrecy Sender.*

"1. With the early form of secrecy sender, and a difference of wave-length of 15 per cent., on an official test five words were read by the intercepting station, distant five miles from sending station, out of 360 words.

"2. Great improvements have since been made. It is not believed that any other system can read the messages, no matter how close the intercepting station is.

"3. Guarantee. With our latest form of secrecy sender, which we propose to furnish, the variation in wave-length is guaranteed to be only ¼ per cent. between spaces and dots. We guarantee that no other system can read the messages.

D. *Anti-Atmospheric Device.*

"1. Messages have been read by the use of the interference preventer when other systems were entirely unable to read on account of atmospheric disturbances. (Official report.)

"2. It is difficult to express a guarantee in this matter, but it may be stated that in practice when the atmospheric is so strong that the noise can be heard with the telephone held six inches from the ear; when the anti-atmospheric device is connected no atmospheric can be heard at all, and the signals are not weakened at all.

"With the forms of apparatus used during the past year, atmospheric disturbances so strong that they can be heard with the telephone six feet away from the ear, are cut down by the use of the atmospheric disturbance preventer so as to be inaudible when receiving, and therefore not to interfere at all with the reception, and this without the strength of the signals being weakened.

E. *Intensity Regulator.*

"This is a device for modifying the intensity of the emitted waves without changing their wave-length, for the purpose of communicating with nearby ships.

F. *Wave Measurer.*

"1 and 2. This device is used for measuring wave-lengths, and is believed to be accurate to about $\frac{1}{4}$ of 1 per cent.

"3. We guarantee this device to be capable of detecting difference of wave-length of $\frac{1}{4}$ per cent., and to be capable of measuring wave-lengths at a distance from the sending station."

U.S.S. KEARSARGE, 1st Rate,
BAR HARBOUR, MAINE, *6th August* 1905.

SIR,

1. I have the honour to make the following report on the test of the Fessenden Wireless Telegraphic apparatus between the U.S.S. "Alabama" and U.S.S. "Illinois" on the afternoon and night of the 2nd inst., and forenoon of the 3rd.

2. The apparatus was handled, both in preparation for, and during the test by the regular ship's operators. In obedience to orders from the Commander-in-Chief, I reported on board the "Illinois," and observed the working of the apparatus on that ship. The test took place at sea, over clear water between the latitudes of 38° and 44° north, and longitudes of 68° and 70° 48′ west, and lasted from 1 P.M. of the 2nd inst., until 3.30 P.M. on the 3rd.

3. The "Illinois" separated from the Fleet at Nantucket Shoals Lightship at 12.30 P.M., 2nd August, and steamed on a course south-west by south at a speed of twelve knots. The "Alabama" continued with the Fleet, and steamed generally on a course north-east by north, at a speed of from ten to twelve knots. In accordance with the prearranged schedule, the "Alabama" sent her position in latitude and longitude on the hour and

thirty minutes, and the "Illinois" sent hers on the fifteen and forty-five minutes.

4. Sending with a normal strength of from thirty to thirty-five amperes, fifty to sixty volts on transformer, and with a spark-gap on each side of the cooling plate of from $\frac{3}{16}$ inch to $\frac{1}{4}$ inch, the ships continued in easy communication with each other up to 11 P.M., at a distance of 230 miles, when the "Alabama" asked the "Illinois" to send stronger, increased spark-gap on "Illinois" to $\frac{5}{16}$ inch each side of cooling plate, and sending current to forty amperes. The "Alabama" continued to receive up to midnight, a distance of 251 miles, after which she failed to make out any message from the "Illinois," or even satisfactory parts of messages, until the following morning at 9.45, when she got part of the "Illinois'" position message, but was interrupted by the Maryland and Boston Navy Yard. The distance between the two ships at that time was 267 miles. The reception on the "Illinois" continued good, and the position of the "Alabama" was correctly received for mid. and 12.30 P.M. The position for 12.30 was sent up to 12.45, and the distance between the two ships at 12.45 was 268 miles.—Very respectfully,

J. M. HUDGINS,
Lieutenant, U.S. Navy.

THE CHIEF OF BUREAU OF EQUIPMENT,
Through the Division Commander, 2nd Division, and the Commander-in-Chief, North Atlantic Fleet.

"DEPARTMENT OF THE NAVY
BUREAU OF EQUIPMENT,
WASHINGTON, *28th November* 1904.

"TESTS OF THE FESSENDEN SYSTEM.

"91. On 15th August the representatives of the National Electric Signalling Company began to install the Fessenden instruments at the Navy Yard station, and on the 19th at the Highlands.

"(92-99. Description of rules governing the test, one rule being as follows:—'Absolutely no messages or signals to be sent by either station during the time assigned to the other station—not even an acknowledgment.')

"100. At 10 P.M. the Board adjourned until 10 A.M., 30th August, when it met to make the interference test; Captain Rodgers and Lieutenant Edgar at the Navy Yard station; Commander Peters and Commander Fiske at the Highlands station; Lieutenant-Commander Jayne on board the 'Topeka.' From 10.23 to 10.57 a press despatch of 861 words was received from the Highlands at the rate of twenty words per minute, with the loss of two words in the headline and two mistakes in the text. During this time, until 10.45, with the exception of intervals of a few minutes, the 'Topeka' was interfering continuously. The Topeka' was anchored off Tompkinsville, five and a half miles distant from the Navy Yard station.

"101. From 11.27 till noon the receiving ship 'Hancock' at the Navy Yard, 630 yards' distance from the Navy Yard station, and the experimental station in building §75 at the Navy Yard, 226 yards from the Navy Yard station, interfered continually. During this period the reception from the Highlands was not good, one message being received with the loss of one word in three.

"(*Note.*—The operators were not notified that this test would be made, and were not prepared with apparatus to cut out nearby interference, hence the loss of one word in three. During the dinner hour an improvised arrangement was fixed up, the results with which are given in paragraph 102. It is evident that the one word lost in every twelve could have been obtained by repeating the message, but no repetition was allowed.)

"102. At noon the Board took a recess until 2.15, then met and continued the test for interference. The station at building §75 interfered from 2.15 until 3 P.M. During this period the average loss was one word in twelve, in a message sent from the Highlands. From 3.15 to 3.45 the secrecy sender was tested. A message of 362 words was sent from the Navy Yard station to the Highlands by a Navy operator, at the rate of thirteen words per minute (Continental Morse), with a secrecy sender. The 'Topeka' had instructions during this period to read the message if possible.

"103. [Relates to time Board adjourned.]

"104. [Gives names of operators.]

"105. At 3.25 P.M., 29th August, Professor Fessenden sent a message to Philadelphia for his operator at that station to come to the New York Navy Yards; the message was acknowledged at once, and the operator appeared at the Navy Yard station at about 7.30 P.M. Both of these messages were read at the Highlands. He brought with him record showing that he had received, in Philadelphia, messages sent from the Navy Yard to the Highlands, one of which was a difficult code message.

"106. The Slaby-Arco apparatus was used aboard the 'Topeka.'

"107. The U.S.S. 'Hancock,' with improvised instruments, used a maximum spark of $\frac{1}{4}$ inch, 15 amperes, and 118 volts. Length of aerial, 120 feet.

"108. Building §75 used improvised instruments, spark $\frac{1}{10}$ inch. Length of aerial, 120 feet.

Address—Bureau of Equipment, Navy Department.

WASHINGTON, D.C.,
7th April 1905.

GENTLEMEN,

1. Referring to your communication of the 5th inst., requesting information as to results obtained with your apparatus by Lieutenant Hudgins, the Bureau quotes from that officer's report, as follows:—

(*a.*) "The interference preventer will practically exclude any wave varying in length by as much as 3 per cent. from the wave to which the preventer is adjusted, provided the intensity of such interfering wave is not greater than that of the wave to be received. As the interfering wave increases in relative intensity, the difference in wave-length must be increased also. Unless very accurately adjusted, the interference preventer absorbs part of the strength of the wave, and the more imperfect the adjustment the greater the absorption. The adjustment of the interference preventer is somewhat difficult, but once made it is practically constant.

(*b.*) "With this connection, it is possible to absorb or shunt to earth waves so strong that close connected they will burn out the receiver, and at the same time to build up a weak resonant wave to sufficient strength to be read. A variation of 10 per cent. in wave-length has been found sufficient to cut out waves so strong that they could not be received direct either on the electrolytic or coherer receivers, while at the same time a weak wave which the coherer would just record, was plainly received on the telephone. With waves of more nearly equal intensity, a variation of 3 per cent. in length is ample.

(*c.*) "With atmospheric or other interference the electrolytic receiver is the better. The sharp degree of tuning practicable with this receiver, permitting the tuning out of other stations, unless very close to the same wave-length, or else much stronger than the wave which it is desired to receive. Atmospheric disturbances pass through the electrolytic receiver with a distinctly characteristic sound, and, unless very numerous or strong enough to burn out the fine point in this receiver, do not materially interfere with the reception of a message coming in at the same time. —Very respectfully,

"H. M. MANNING,
"*Chief of Bureau of Equipment.*"

"National Electric Signalling Co.,
8th and Water Streets, Washington, D.C.

"U.S. NAVY WIRELESS TELEGRAPH STATION,
NEWPORT, RHODE ISLAND, 24*th March* 1905.

"This is to certify that the interference preventer, as arranged by Mr Chas. J. Pannill of the National Electric Signalling, to my personal observation cut out the following stations absolutely:—Cape Cod, Fort Wright, New Haven; Marconi boats as well as their Siasconsett Station; Jersey City; the Fall River boats, 'Priscilla' and 'Pilgrim' (about a half hour after sailing from Newport); moderate atmospheric, also terrific atmospheric, reduced about 50 per cent., and received Nantucket Shoals Lightship, without decreasing her signals, but on the other hand increasing them as compared with receiving them on [another] company's apparatus installed at this station. All of the above stations, excepting Jersey City, which we do not get on the other apparatus, interfere with our receiving from Nantucket Shoals Lightship.

"On Saturday, 18th March, I received signals from the Fessenden Jersey City Station, which were very distinct.

"The atmospheric disturbances were so severe one night last week I was not able to receive the report of a passing steamer from the Lightship on the —— apparatus, but upon changing over to the Fessenden interference preventer, the terrific atmospheric was cut down to such an extent that I was enabled to receive the report without any trouble.

"A. Y. FORREST,
"*Electrician, 2nd Class.*"

First Successful Exchange of Wireless Messages across the Atlantic.

"Towards the end of December 1905 the National Electric Signalling Company of Washington, U.S.A., completed its two Transatlantic stations, one of which is situated at Brant Rock, Massachusetts, about twenty miles south of Boston, and the other at Machrihanish, Kintyre, Scotland, the distance between the stations being a trifle over 3,000 miles.

"The power is developed by a boiler-engine-alternator equipment having a maximum capacity of 25 K.W. 60 cycles current. A transformer steps up the voltage to about 25,000, thus charging the condensers, which are discharged by means of a gap adjustable so as to effect the discharge at any desired point of the cycle.

"The receiver used is the liquid barretter in its latest form. The aerial is formed by a tower extending to a height of 415 feet above the ground level, and supporting a sort of umbrella formed of wires at its top. The tower is essentially a steel tube 3 feet in diameter, supported every 100 feet of its height by a set of four steel guys, there being thus sixteen guys in all. The tower is pivoted at its base on a ball and socket joint, and is insulated from the ground for a voltage of 150,000. The guys are insulated from the tower as well as from the ground, besides they are divided into 50-feet sections by means of strain insulators. One of the most serious problems to be solved was the construction of these strain insulators, which, while capable of safely transmitting the maximum stress of about 20,000 lbs., also resist an electrical tension of 15,000 volts each. The maximum deflection of the top of the tower in a 90-mile hurricane is computed to be 15½ inches. A wave chute containing over 100 miles of wire, and extending over six acres, is a very essential feature of the installation.

"On 3rd January 1906 the first signals were received from the American side, and shortly afterwards communication was established, so that messages were freely exchanged at night. The intensity of the signals received by telephone was at times so great that messages could easily be read with the diaphragm three inches from the ears of the operator. A station twenty miles distant from Brant Rock, using about

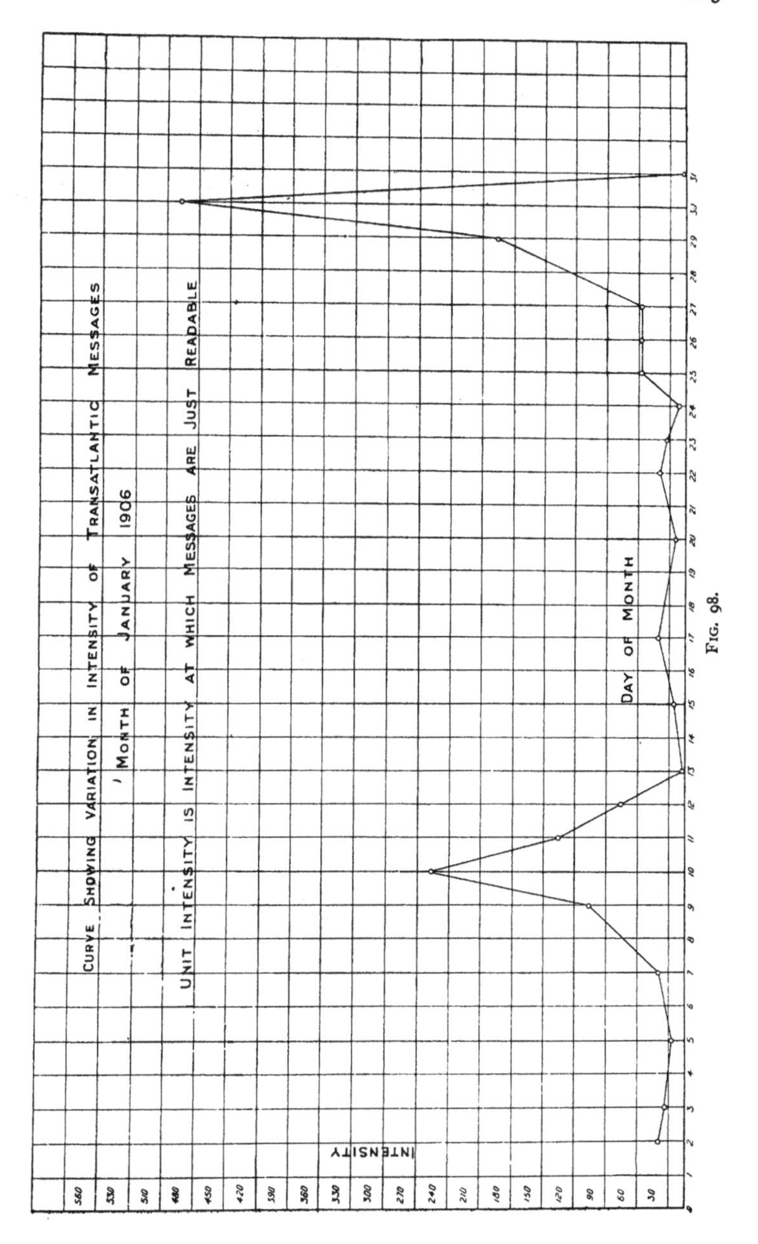
Curve Showing Variation in Intensity of Transatlantic Messages
Month of January 1906
Unit Intensity is Intensity at which Messages are Just Readable
Day of Month
Intensity
560 530 510 480 450 420 390 360 330 300 270 240 210 180 150 120 90 60 30
0 1 2 3 4 5 6 7 8 9 10 11 12 13 14 15 16 17 18 19 20 21 22 23 24 25 26 27 28 29 30 31

FIG. 98.

30 K.W., and sending on a wave-length differing not much more than 3 per cent. from that of the Scottish station, was cut out while messages were received from Machrihanish.

"The company is at present endeavouring to establish continuous night and day communication. With this in view alternators of very high frequency will shortly be installed at both stations, together with the apparatus accessory thereto."

CHAPTER XI.

THE HOZIER-BROWN SYSTEM.

Distinctive Features of the System.—Colonel Sir H. M. Hozier, and Mr S. G. Brown have kindly supplied me with the following notes on their system :—

"The 'Hozier-Brown' system of wireless telegraphy does not vary from most of the other systems except in the use of a special receiver, and in the method of recording the messages received.

"The transmitter, for moderate distances up to, say, sixty miles, consists of the usual spark coil, an aerial wire of, say, 150 feet in height, and a Morse key to interrupt the current through the primary of the spark coil so as to transmit the message according to the Morse code. The receiving aerial would be of the same height, viz., 150 feet.

"The receiver, or detector, an invention of Mr S. G. Brown, consists of a pressed pellet of peroxide of lead placed between a platinum plate and a blunt lead point; the blunt lead point is carried on the end of a steel spring, and is capable of adjustment by means of a milled headed screw to press more or less lightly upon the peroxide.

"The theory of the detector is as follows :—

"The detector exerts a certain back E.M.F., which does not vary with the number of cells used in its circuit. A 2 volt accumulator in its circuit will be found to give the best results. The lead peroxide behaves as an electrolytic cell when in conjunction with its electrodes of lead and platinum. Similarly to electrolytes lead peroxide has a negative coefficient of temperature and resistance.

"The action of the normally flowing current from the storage cell is illustrated in Fig. 99. Ions of Pb (+) tend to pass upward to the lead cathode; ions of O_2 (−) downward to the platinum anode. While the detector is unaffected by Hertzian waves, however, this tendency is neutralised by the opposing tendency of the Pt. PbO_2. Pb combination to act as an independent battery sending a current in a direction the reverse to that of the external current, sending ions of O_2 to the lead and ions of $\overset{+}{Pb}$ to the platinum electrode. Under the influence of the alternating current resulting from the Hertzian waves, this tendency of the combination to act as a battery is brought actively into operation, with the result that the effective

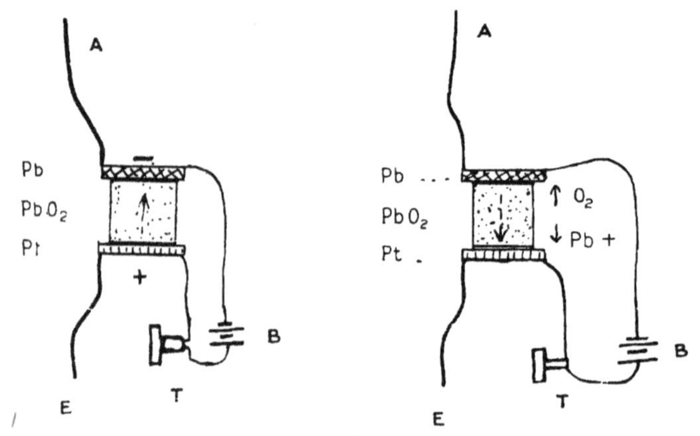

FIG. 99.—HOZIER-BROWN RECEIVER.

A, Aerial Wire; *E*, Earth; *B*, Battery of about 2 volts; *T*, Telephone Receiver. The detector consists of lead peroxide between lead and platinum plates, its voltage is normally in opposition to that of local battery, and nearly balances it (see I.). The received jigs cause action shown in II., which destroys balance of opposing E.M.F.'s and causes current audible in telephone.

current sent by the external cell is diminished. When the radiation intercepted ceases, the permanent current removes the films of lead (from the bottom) and of oxygen (from the top).

"That this explanation suffices, to some extent at least, is shown by the fact that if a metal (such as iron) more electro-positive relatively to platinum than lead, be used for the upper electrode, the deflection of the needle due to the first waves received is more considerable than in the case of lead, although the combination becomes less and less sensitive to further radiation, while the 'after-creep' of recovery is very much slower than in the case of lead owing to the fact that electro-positiveness and ease of oxidation go together, and it is only very slowly that the permanent current can remove the film of oxide formed upon the iron electrode. The detector is extremely sensitive to Hertzian waves, and being dry has very little electrostatic capacity.

"As stated above, the platinum, peroxide, and lead point act as a battery when under the action of the Hertzian waves, and if joined in opposition to the external 2-volt battery, will cause the current flowing through it and round the circuit to drop in value. The detector does not act as a battery when the Hertzian waves cease, and the external battery is thus enabled to increase the current flow. It will thus be seen that the current diminishes in value when the Hertzian waves act on the detector; that is to say, the detector acts as an anti-coherer.

"The detector is quite automatic in its action. The current decreases in value on the receipt of the signal, and increases instantly in value when the Hertzian waves cease: the amount of current variation is more or less proportional to the strength of the Hertzian waves.

"There are three ways of showing or indicating the received signals. One is to place a sensitive voltmeter in the circuit, and it is not uncommon to see a variation of over half a volt registered for each signal received from a transmitter worked over fourteen miles away. A telephone in the circuit will indicate by the note the long and short signals of the Morse code sent from the distant station, and when the Hertzian waves are strong the noise in the telephone may become uncomfortably loud, while signals sent over 100 miles can usually be heard quite distinctly. The third method is to place a *siphon direct writer* in the circuit.

"The current from the detector flows round the suspended coil of the direct writer. The suspended coil hangs in a strong magnetic field, and carries a glass siphon tube; one end of the tube dips into the ink-well, the other end bears upon the moving strip of paper. On the arrival of the signals the suspended coil is deflected to one side either for a long or short space, to indicate the long or short signals of the Morse code; and thus the message is recorded on the strip of paper. (The suspended coil usually carries in addition a relay contact arm to bring in the current from a local battery, and ring a bell as a warning to the operator that a message is arriving.)

"It will be seen that (with the exception of the ringing of a bell) the message is written or recorded without the employment of any relay whatsoever, the amount of current delivered by the detector being ample to record large signals even with a moderately insensitive direct writer.

"The advantages of our detector are great sensitiveness together with great current changes, allowing the recorders to be operated directly without the employment of any relay. The detector is moreover extremely robust and cannot easily be damaged: the instrument is self-operating, works with extreme rapidity, will record as fast as any operator can send, and is practically everlasting, as it is very rarely that the peroxide pellet need be changed. As regards sensitiveness, on a 185 feet aerial signals have been received and recorded over 100 miles; this, it must be remembered, with the detector coupled directly to the aerial and to the earth-plate.

"No receiving transformer is used, the detector being coupled direct to the aerial, as before stated.

"The chemicals of the detector are carried in a brass case, together with a switch, so that on *sending* in the station, the chemicals can be disconnected from the outside circuit by turning the switch and completely isolated."

Production of Continuous High Frequency Currents for Use in Transmission.—Mr S. G. Brown has recently patented a method of producing uniform current jigs, of which the following is a description:*—

"The positive electrode, as will be seen from the figure, consists of a disc of metal, preferably aluminium, which is kept slowly rotating, the negative electrode consisting of a copper block resting against the edge of the disc. On the

* Given here by the kind permission of Mr Brown and the proprietors of the *Electrician*.

passing of a direct current at 200 volts pressure through a resistance and a high self-induction, and through the disc to the copper block, exceedingly rapid oscillations were produced in a capacity shunting the moving contact.

"Measurements of the current flowing in this condenser showed by calculation that the oscillations were in some instances of the order of millions per second, and a small incandescent lamp consuming 1 ampere at 20 volts could be brilliantly lit up through a condenser of 0.005 or less of a microfarad, although the effective E.M.F. across the

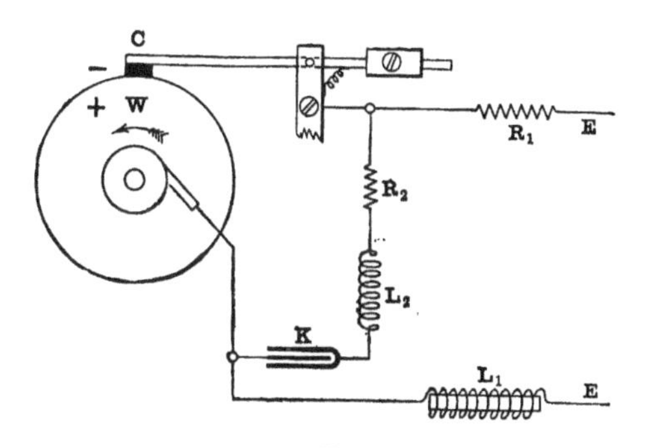

FIG. 99A.—BROWN'S UNIFORM JIG PRODUCER.

W, rotating disc; *C*, balanced copper block; *E*, *E*, direct-current electric light leads; R_1, series regulating resistance; L_1, inductive coil of, say, 20 henries to steady current; *K*, capacity shunting contact on disc; R_2, resistance; L_2, air core inductive coil to tune circuit, although this coil is not necessary to produce oscillations through condenser.

If voltage is higher than, say, 200 volts, two or more discs and copper blocks may be employed as a multiple spark-gap.

electrodes could not, from calculation, have exceeded 50 volts.

"Many experiments with this arrangement were tried, which fully proved the existence of these high oscillations, and their continuous production; for instance, it was possible to light up an incandescent lamp through a 9-in. brick wall, employing only two loops of wire of one single turn, each enclosing an area of some 3 sq. ft., and placed on either side of the wall, the oscillations flowing in the one loop and the lamp being in the circuit of the other.

"The advantage of this moving disc was that it was self-cooling and self-regulating, this self-regulation being effected by fixing the copper block on one end of a counter-balanced arm, so that it could be adjusted to float with a definite pressure on the edge of the disc. This instrument required little or no attention, and has been worked all day without adjustment and with no appearance of change taking place."

CHAPTER XII.

WIRELESS TELEGRAPHY IN ALASKA.

ONE of the most remarkable achievements in practical commercial wireless telegraphy is the installation and maintenance by the United States Army Signal Corps of wireless communication across Norton Sound, in Alaska.

Norton Sound is a great bay at the mouth of the Yukon River, 107 miles wide between the points where the wireless stations stand. One of these, St Michael, on the southern side of the bay, is the terminus of the land lines which run 1,300 miles southward to Fort Liscum, which is in communication with Seattle by a cable 1,600 miles long; the other, on the northern side, is at Safety, near Nome, the most important centre in the north-western district of Alaska. It had been found impossible to maintain a cable across the gulf on account of the ice, and the absolutely barren and snow-swept nature of the coast made an ordinary land line round it almost impossible. As Nome is the centre of a large fishing and gold-mining district it was essential to provide telegraphic communication, especially as the usual means of communication are only available during a very few months in the year.

The United States Army therefore decided, after three years had been wasted in the failure of some commercial companies to provide a service, to establish wireless communication through the agency of the Signal Corps. Captain Leonard D. Wildman, who designed the stations, and superintended their erection and working, has most kindly favoured me with drawings and descriptions, from which the following account is taken.

I make the following quotations from the Annual Report of the Chief Signal Officer, U.S. Army, for 1905, in order to compare the reliability of land lines and wireless telegraphy in these regions. This officer writes :—

"The conditions considered, the continuity and reliability of the service (on the land lines) has been beyond expectation, the interruptions between Valdez and Nome, the extremes of the line, being 40.7 days, or one day in nine. . . . Half of the interruptions occurred in June owing to forest fires in the upper Tanana, and unprecedented flood conditions in the lower valley. This unusually efficient service is the result of devoted field work by the Signal Corps, assisted by men of the line. . . . The interruptions during the winter were reduced to about one-half of those of previous years, due to the energy and zeal of the repair men who patrolled the line on their section."

Contrast this with the above :—

"In August 1903 a wireless section of 107 miles across Norton Sound was established through the professional skill and exceptional ability of Captain Leonard D. Wildman, Signal Corps. This is the only long wireless telegraph system in the world, it is believed, that is regularly operated as a part of a system handling commercial business. On 6th August 1904 it completed a year of uninterrupted service over its course of 107 miles. It has handled *daily and uninterruptedly* the entire telegraphic business of Nome and the Seward Peninsula, which, together with the official business, averages several thousand words daily. More than a million words were sent during the year, many thousand being commercial code words, in which no error has ever been traced to this section. In a single hour there have been transmitted over this section 2,000 words without error or repetition. The successful installation and operation of this unique work by the officers and soldiers of the American Army afford an added illustration of their intelligence and aptitude."

It is certainly remarkable that this unique record should have been accomplished at stations, one of which was built upon a glacier, and both on a coast only accessible during about three months each year. It reflects great credit both on the designer of the stations and on those who assisted him in their working. To continue quotation :—

"The wireless work was done by an alternate current, 500 volts, 60 cycle, 3 kilowatt generator; a 6 horse-power gasoline, single cylinder engine, with special governor; a small grid at masthead with only two connecting wires, and large ground capacity. The dynamos, built to Wildman's plans, were provided with specially heavy insulation about the armature coils and collector rings. The switchboards were of home manufacture.

"Electric storms gave little trouble, and weather or ice conditions had no material influence. The receiver was of the De Forest type (now declared to be covered by Fessenden's patents), modified by Signal Corps inventions. Captain Wildman found many opportunities for the resourceful minds of himself and his subordinates. Broken Leyden jars were successfully replaced by air condensers, the spark was muffled, currents shunted, and many other improvements applied. Captain Wildman thinks duplexing possible to a certain extent, and a call-up device probable. . . . Sergeant M'Kinney devised a key that increased the sending capacity from 15 to 30 words per minute."

So much for the success of the system; to go now into details. It seems most probable that Captain Wildman's system of tuning the aerial to the closed oscillating circuit, as shown in the American patent 764,093, dated 5th July 1904, was the first use of the common adjustable inductance with the hot-wire ammeter in the antenna to indicate maximum resonance. It is certainly the simplest way, and has been adopted, either independently or following his idea, by a number of companies.

The use of a practically wattless current in the primary circuit when the key is open is a novel and valuable feature of this system. The means for obtaining it are similar to Professor Fleming's inductive shunts on transmitting keys, but with this difference that Professor Fleming's "choking coil will stop all currents from flowing into the transformer;"* while Captain Wildman adjusts the inductance to give a nearly wattless current in the primary. The result as proved by practical experience is a considerable reduction in the power used, and consequent saving in gasoline. This is an important economy in outlying stations.

Another feature of the system is the use of two earth connections with proper inductances, one of which goes permanently through the receiver, and the other, with a small extra spark-gap in it to obviate the necessity for a switch, is used for transmitting. The arrangement was subsequently re-invented, no doubt independently, and patented by the Marconi Company.

An improvement which adds to the comfort of the operator is an arrangement whereby the main spark-gap is put outside the house. For tuning purposes there is an alternative gap inside which is not used during actual transmission.

The outside gear shows some ingenious modifications particularly useful in the stormy climate in which they were devised. The aerial, for instance, consists of two wires attached to a small grid at the top and a cross stretcher at the bottom. A rope is also fixed to the top grid, and hanging between the wires helps to bear the weight of ice which accumulates on them in winter. The lower end of the rope carries a weight which hangs clear of the ground. The whole aerial is therefore like a heavy pendulum, and is brought back to its vertical position by the weight after being deflected by a gust of wind.

We shall return to Captain Wildman's experiments on the effect of wind and moisture on transmission in a later chapter.

---

* "Electric Wave Telegraphy," p. 504.

CHAPTER XIII.

THE DE FOREST SYSTEM—THE POULSEN SYSTEM.

THESE two systems, though differing in character, may be conveniently referred to in a single chapter.

De Forest System.—I have already mentioned several of Dr De Forest's inventions in earlier chapters; it is, therefore, only necessary to say that, in common with other systems of the same class, an aerial wire, or network of wires, of considerable height is used.

Through the kindness of the Amalgamated Radio-Telegraph Company Limited, of London, I am enabled to add a note on the performances of some stations fitted with the De Forest apparatus.

One of the largest of the stations of this system is that situate at Manhattan Beach, the property of the American De Forest Wireless Telegraph Company, which has established communication with Colon (2,170 miles), and with Porto Rico, Boston, Montreal, Cleveland, Chicago, and St Louis.

The Manhattan station is used for distributing news collected in New York City to radio-telegraph stations within many hundreds of miles. More than ninety stations have been built by the American De Forest Wireless Telegraph Company, while many others are under construction. Illustrations of two of the most important stations, those at New York and Boston, are given in Figs. 100 and 101.

The De Forest patents for the whole world, with the

FIG. 100.—NEW YORK STATION OF THE DE FOREST WIRELESS TELEGRAPH COMPANY.

exception of the United States of America, Canada, and the Bermudas, are now the property of the Amalgamated Radio-Telegraph Company Limited.

Poulsen System.—This system constitutes a remarkable advance in wireless telegraphy. Primarily, the generator

FIG. 101.—BOSTON STATION OF THE DE FOREST WIRELESS TELEGRAPH COMPANY.

is, in principle, based upon the singing arc of Duddell. A direct-current arc is shunted by a condenser-circuit in which alternating currents are generated. In the case of the Duddell arc, it is only possible to generate efficient oscillations in the shunt-circuit when the oscillation factor of the latter is large, and the natural periodicity, therefore, low. Thus, although the singing arc gives continuous, undamped oscillations, these are not sufficiently rapid to serve the purposes of wireless telegraphy.

Mr Valdemar Poulsen, after a long series of experiments, has discovered a method of producing continuous oscillations which are not only of high frequency, but also of great efficiency. The chief distinction between the Poulsen generator and the Duddell arc relates to the use of hydrogen, or an atmosphere containing hydrogen, as the surrounding medium of the arc employed. As the result of Poulsen's discoveries, we can now generate powerful oscillations having a frequency of hundreds of thousands per second, whereas formerly such oscillations could only be obtained at acoustic periodicities.

The receiver is the invention of Mr Pedersen, and is designed to take full advantage of the power of the continuous series of impulses to create revibration in an equifrequent circuit. The oscillating circuit of the receiver is only intermittently connected to the detector. Thus the jig has time to grow without hindrance in the complete circuit through superposition of the waves of alternate current received, the detector being only introduced when a large current jig has been induced.

The Amalgamated Radio-Telegraph Company Limited, of London, has acquired the Poulsen patents for the whole world, with the exception of the United States of America.

A full description of this system was given in the *Electrician* for 16th November 1906.

CHAPTER XIV.

THE TELEFUNKEN SYSTEM.

THIS system is based on the patents of Professors Slaby and Braun and of Count von Arco. It may be considered as striking a mean between the earlier systems of Marconi and Lodge-Muirhead, though of course with many variations

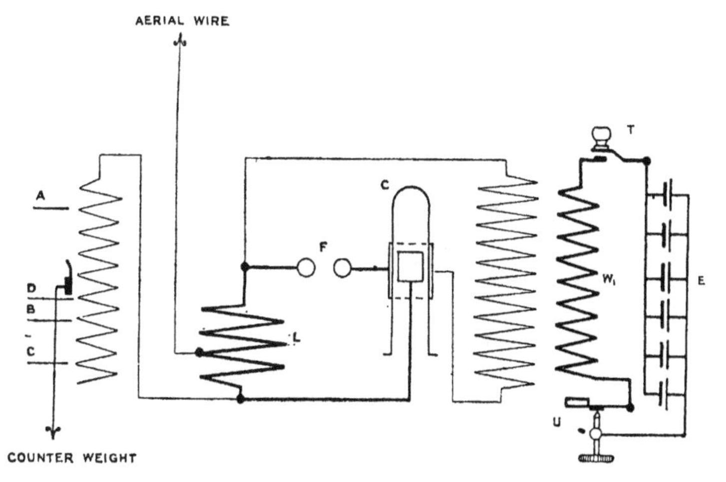

FIG. 102.—TELEFKUNKEN TRANSMITTER.
E, Battery; *T*, Key; *W*, Primary of Induction Coil; *U*, Interrupter; *C*, Capacity; *F*, Spark Gap; *L*, Inductance; *A*, Variable Inductance.

and elaborations in detail. The system has made very rapid strides in its evolution, owing largely to the encouragement given to research in Germany. A system of wires, similar to those used by Marconi, forms the aerial, and the earth connection is given by a large capacity, as in the

Lodge-Muirhead apparatus. A coherer and receiving circuit in many respects similar to the Marconi arrangement is employed. The annexed figures, kindly lent by the Telefunken Company, show the general arrangement of the apparatus.

Owing to good commercial and technical management,

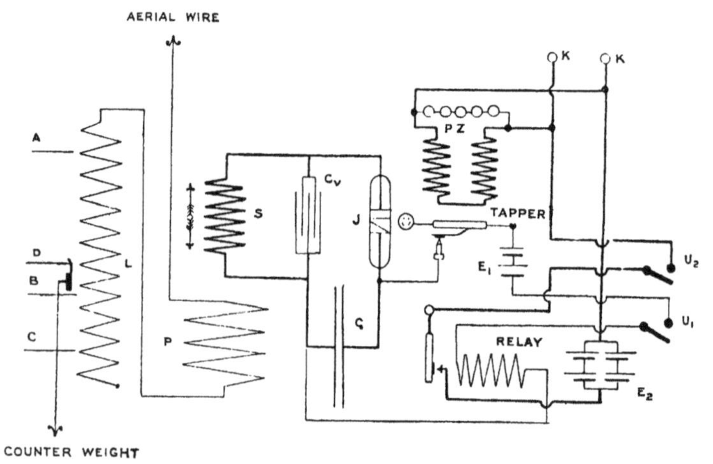

FIG. 103.—TELEFUNKEN RECEIVER.

P, Primary of Receiving Transformer; *S*, Secondary of Receiving Transformer; *C*, *Cv*, Condensers; *J*, Coherer; E_1, Coherer Battery; E_2, Tapper and Inkwriter Battery.

assisted by the powerful influence of the German Government, this system is now probably the largest in the world, 518 stations having been erected before the close of the year 1905.

For further details of this important system the reader is referred to "Wireless Telegraphy" by Gustav Eichorn.

CHAPTER XV.

DIRECTED SYSTEMS.

THOUGH not invariably advantageous, it is sometimes useful to be able to limit the transmission to approximately one direction. Thus a large number of stations at different points might communicate with one another in pairs without interference or the necessity for tuning. It would also occasionally be very useful, especially for maritime purposes, to be able to determine the direction of the station from which a message has come. The two problems are distinct, the latter being, at present at least, the more important. If a ship were able to tell the bearings of a land station from which she is receiving a message it would be an assistance to navigation in thick weather ; while, if the bearings of two coast stations could be found simultaneously, the actual position of the ship would be easily calculable by trigonometry. It would not even be necessary that the actual determination of the bearings of the station should be made on board the ship. The bearing of the ship from each station might be taken at the stations and telegraphed out to the vessel. Thus it is not necessary for this purpose that a vessel should carry any special apparatus beyond the ordinary instruments required for non-directed wireless telegraphy. The special apparatus can more conveniently be kept at the stations on the coast.

With Mr Marconi's first system, Hertzian waves were used, which were reflected from parabolic mirrors, as in Hertz's experiments, and of course the radiation was directed as a ray from the mirror. So well does the mirror achieve the purpose of direction that it becomes difficult to maintain communication between the stations, as the slightest movement of the reflector so alters the aim of the ray that it may pass above or below the receiving station without touching it. Wireless telegraphy of this kind would be almost useless at sea on account of the difficulty of keeping the ray directed towards the receiving station. The ray is invisible, and signals can only be received if it strikes the receiver directly. In this it is unlike the beam from a searchlight, which produces readable signals, even though the ray does not strike directly the eye of the observer, but is thrown on a cloud above him. Two vessels can thus communicate by searchlight signalling, though they are invisible to one another below the horizon.

It would thus be very useful to be able to tell the direction from which a signal comes, but of much less general importance to be able to limit the transmission to a given small angle.

Many attempts have been made to obtain directed waves, most of them depending on the expected interference of the waves from two aerials placed half a wave length apart. If the jigs in both were in the same phase, it would appear that the radiation should be zero in a vertical plane through the wires and a maximum in a vertical plane at right angles to this. In 1899 the author commenced experimenting with this idea, but extraneous circumstances prevented the experiments getting beyond the preliminary stage. This was probably the first attempt, however, to obtain directed waves by means of aerial wires, and without the use of reflectors. More recently, the subject has been studied, and patents taken out by Braun, Artom, Stone, and many others. Professor Braun's work on phase differences between high frequency circuits, undertaken to obtain data for a method of this type, is a masterpiece of experimental and mathematical skill.* Though presenting no

* *Electrician*, May 25, 1906, and many earlier dates.

great difficulties where the desired direction of transmission is a fixed one, this method would have been impracticable for maritime purposes, particularly in cases where it was advisable to use a wave of considerable length, owing to the excessive size of the apparatus. To find the direction from which the radiation has come would on this system be even more troublesome, as it would involve either a number of fixed aerials, or a fixed one and another movable, so as to accommodate both change of direction and difference of wave-length.

A simpler method * for attaining the same end was found by Mr Garcia in 1900, who used horizontal or inclined aerial wires earthed at one end. These were arranged, so that the effective radiation only took place throughout a small solid angle, and could therefore be directed by altering the direction of the aerials. In a similar manner the direction from which the current was coming could be determined.

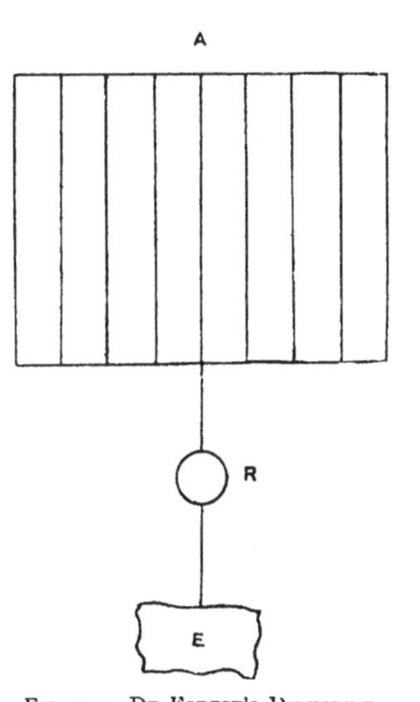

FIG. 104.—DE FOREST'S DIRECTIVE AERIAL, VERTICAL FORM.

A, Grid of Vertical Wires; *R*, Transmitting or Receiving Instrument; *E*, Earth.

In patents applied for on 28th May 1904, Dr de Forest describes two methods by which the direction of the sending station may be determined. The first involves the employment of a large grid of vertical wires which is suspended, so that it may be turned about a vertical axis. The strongest signals are received when the plane of the grid is at right angles to the line joining the stations, and weakest when the plane of the grid is in this line. By taking a few

* See letter by F. Galliot, *Electrician*, May 18, 1906.

observations with the grid in different positions (orientations), the direction of the line joining the stations can be determined with considerable accuracy.

It was found that by using a grid 15 by 6 feet as aerial, and turning about until a maximum strength of current was observed, it was possible to determine within 10° the direction of a station seven miles off.

For practical purposes this means that if two coast stations about five miles apart were situated favourably with regard to a ship about five miles off shore, they could give the bearings with sufficient accuracy to enable the master to determine his position to within one mile. In foggy weather on a dangerous coast even this degree of accuracy would be of great value, while no doubt, as the method is developed,

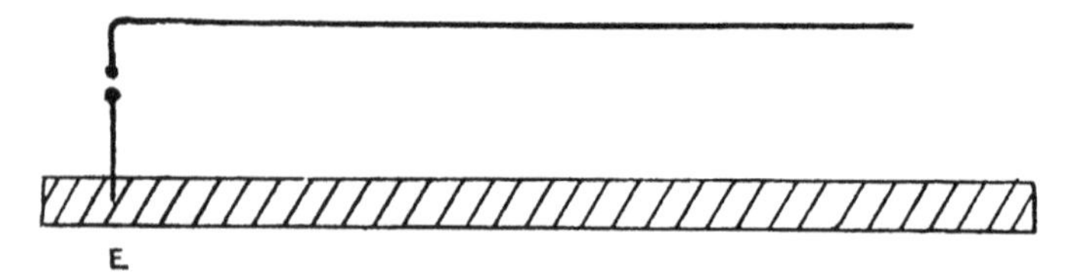

Fig. 105.—De Forest Directive Aerial, Horizontal Form.

very much greater exactitude may be looked for in the future.

In another patent of the same date Dr de Forest describes another method of obtaining the direction of motion of the lines of force concerned in transmission. A long horizontal wire is suspended at a short distance above the ground and has a short earth connection in which the receiver is placed. It was observed that a maximum effect was noticeable when the horizontal wire was an extension of the line joining the stations. Thus by turning the wire in a horizontal plane about its vertical part and noting the position of greatest current, it is possible to find the direction of the sending station.

Either solution of the problem seems applicable directly

to practical working, though possibly the grid would be more easily adapted to existing arrangements, and less cumbersome to handle.

Dr de Forest has also solved the problem of directing the radiation in a similar manner. In March 1901, and more particularly in a patent of January 1904, he has shown that an aerial consisting of a short vertical branch and a horizontal part from the top of the vertical one, sends out most moving lines of force (radiation) on the side of the vertical opposite to the horizontal branch and in the direction of the horizontal wire produced.

Messrs Duddell and Taylor experimented in June 1904* with a transmitting aerial of similar form, and obtained results which, allowing for differences in the transmitting current and a certain want of symmetry in the positions of the earth and aerial, confirm De Forest's results.

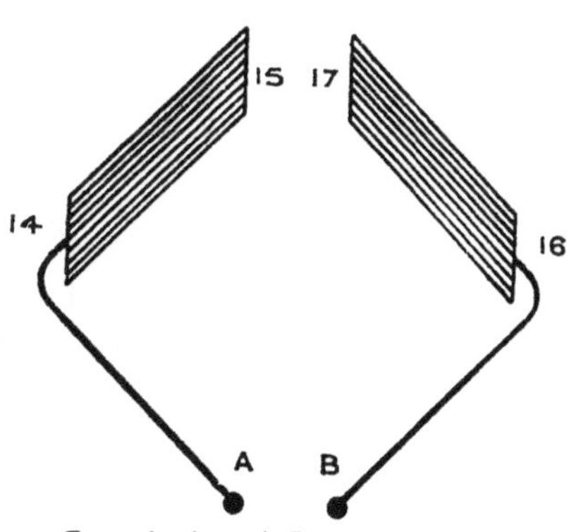

FIG. 106.—ARTOM'S DIRECTIVE AERIAL.
A, *B*, Connections to jig producers giving fixed phase difference; 14-15, 16-17, Elevated Grids.

In July 1905 Mr Marconi patented an arrangement like Dr de Forest's horizontal wires, and in March 1906 read a paper to the Royal Society describing some experiments he had made with the apparatus. The results given are of interest both from practical and theoretical points of view, and have been explained, on the basis of Hertz's theory, by Professor Fleming. S. Artom has also invented a directive aerial, on a different principle, which has given excellent practical results. In a paper published in *L'Elet-*

* *Journal of the Institute of Electrical Engineers*, vol. 35, p. 328.

*tricista** he gives details of his apparatus and results. The aerial used consists of two independent conductors bent so as nearly to form the sides of a square, one corner of which is uppermost. The aerials are attached to a jig generator

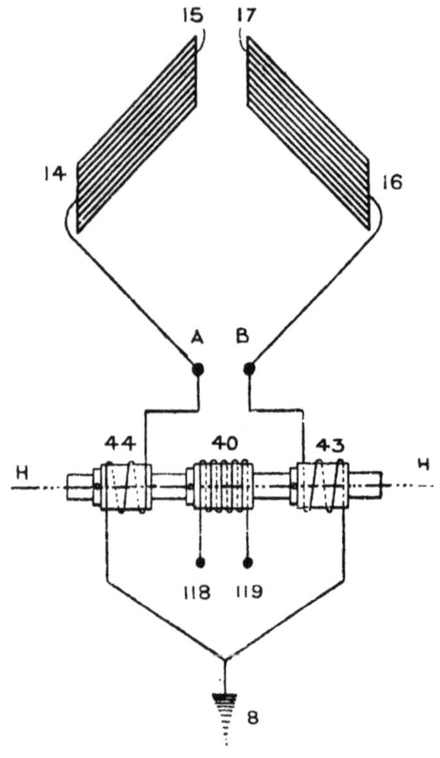

FIG. 107.—ARTOM'S DIRECTIVE RECEIVER.

A, *B*, Aerial Connections; 43, 44, Primary Windings of Receiving Transformer; 40, Secondary Winding of ditto, connected at 118, 119, to Detector; 8, Earth.

which gives currents of different phase. The transmission is a maximum in a line perpendicular to the plane of the aerials and zero in that plane.

* *L'Elettricista*, Rome, July 15, 1906.

The receiving aerial is similar. The receiver itself contains a transformer with three windings, one to each aerial and one to the detector. Unless the proper phase difference exists between the jigs from the aerials, no effect is produced on the third winding, and therefore no signal is received. The system is therefore both directive and selective. It has worked most satisfactorily up to 55 km. between Rome and a number of other stations.

In a former chapter I have defined an electrical jig as a periodic electrical motion of high frequency, such as is generally used in open-circuit wireless telegraphy. In order to understand the peculiarities of such motions it is useful to study the mathematical laws which express their properties numerically. For many reasons we cannot do this at all fully at present. The mathematics of the subject are still in a very rough state, and hardly go beyond a first approximation to the facts. However, since for engineering purposes a first approximation is often good enough, I shall give in the next chapter a few of the more important points in the mathematical theory. More complete discussions will be found in the larger books which have recently been published by various authors.

CHAPTER XVI.

SOME POINTS IN THE THEORY OF JIGS AND JIGGERS.

Properties of Oscillating Motions. — In order that we may start with a clear idea of the nature and properties of an oscillating motion, whether electrical or mechanical, I quote (with the kind permission of the author and the publishers) the following concise description of the simplest form of periodic motion given by Professor Gibson in his "Elementary Treatise on Graphs." * It should be noted that though the motion of a point to and fro in a straight line is all that is mentioned, the distance x of this point from the fixed point which is chosen as origin may represent any single valued quantity such as voltage or current, and thus the one set of equations will serve to represent any simple to-and-fro motion, whether mechanical or electrical. To make the matter still clearer we may suppose that x represents the ordinate of a curve which gives the variations of the electrical quantity with time.

"**Simple Harmonic Motion or Sine Wave.**—When a point is moving in a straight line in such a way that at time t its distance x from a fixed point O on the line is given by the equation

$$x = a \cos (nt + \alpha), \text{ or } x = a \sin (nt + \beta) \quad . \quad . \quad . \quad (1)$$

the point is said to describe a simple harmonic motion.

---

* London : Macmillans.

"The motion is obviously vibratory, or to and fro. The point moves first in one direction to the distance a from O, then back through O to a distance a on the other side, then returns towards O, and so on. The greatest distance from O that the point reaches, namely, a, is called the *amplitude* of the motion.

"As t increases from 0 to $\frac{2\pi}{n}$ (or from t_1 to $t_1+\frac{2\pi}{n}$ where t_1 is any value of t) the point makes one complete to-and-fro motion, $\frac{2\pi}{n}$ is therefore called the *period* of the motion. The reciprocal of the period, namely $\frac{n}{2\pi}$, is sometimes called the *frequency* of the motion. If T is the period and p the frequency, then

$$T=\frac{2\pi}{n};\ p=\frac{1}{T}=\frac{n}{2\pi};\ n=\frac{2\pi}{T}=2\pi p.$$

"The function $a\cos(nt+\alpha)$, or $a\sin(nt+\beta)$, is frequently called a simple harmonic function of t; its graph, that is

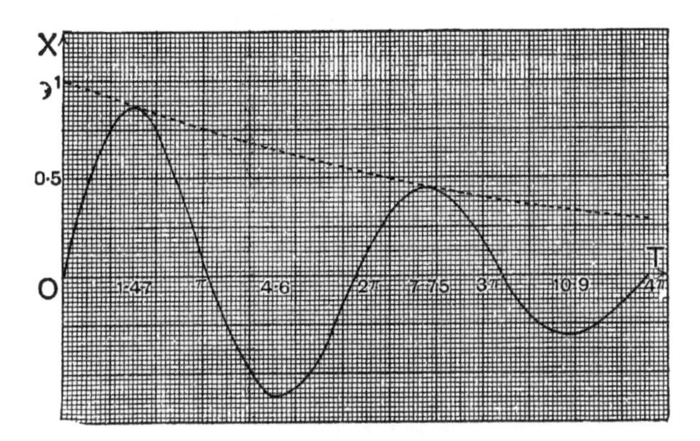

FIG. 108.

the cosine curve or the sine curve, is called a simple harmonic curve. The function is of great importance in all branches of physics.

"The function of t given by the equation (k positive)

$$x = ae^{-kt}\cos(nt+\alpha) \text{ or } x = ae^{-kt}\sin(nt+\beta) \quad . \quad . \quad (2)$$

is sometimes called a simple harmonic function with decreasing amplitude. The co-efficient ae^{-kt} of the cosine or sine is a function of t, which decreases as t increases. Physically the equation represents what is termed a *damped* vibration.

Fig. 108 is the graph of

$$x = e^{-\frac{t}{10}} \sin t \quad . \quad . \quad . \quad . \quad (3)$$

and gives some idea of the nature of the function. Two waves are shown, but after a period of sin t the height becomes very small. Thus, when $t = 10\pi + \frac{\pi}{2}$, we find

$$x = e^{-3.30} \sin \frac{\pi}{2} = 0.037.$$

"The dotted curve is the graph $e^{-\frac{t}{10}}$ which touches the other graph near the crests of the waves; at the first crest $t = 7.75$. The hollows (the minimum values of x) are given by $t = 4.6$ and $t = 10.9$.

"The amplitude of the function (2) when t has any value t_1 is ae^{-kt_1}; when t has increased by $\frac{1}{2}$T (where T is the period of $\frac{2\pi}{n}$ of the circular function) the amplitude has decreased to $ae^{-k(t_1+\frac{1}{2}T)}$. The ratio of the first to the second of these amplitudes is—

$$ae^{-kt_1} : ae^{-k(t_1+\frac{1}{2}T)} \text{ or } e^{\frac{1}{2}kT}.$$

The Napierian logarithm of this ratio, namely $\frac{1}{2}k$T, is called the *logarithmic decrement* of the amplitude."

Turning now to more technical mathematics, we have firstly, as is usual in the theory of electricity, an investigation of Lord Kelvin's to consider. The discharge of a condenser had been studied experimentally by Henry and others when Lord Kelvin attacked the problem mathematically, and gave a solution which accorded entirely with the results of observation. For the benefit of readers to whom mathematical language does not appeal, I may explain that the chief results arrived at are as follows:—

A conductor is supposed, charged, and suspended above the earth by insulating supports. It is then suddenly connected to the earth by a wire. It is shown that a discharge takes place which is unidirectional and slow if the ratio of the inductance of the wire to the capacity and resistance of the system is below a certain value, and oscillatory if this ratio is above the critical value. The period, amplitude, and damping of the oscillatory discharge are calculated, and the amount of work done, on the supposition that there is no radiation of electrical energy, but that all is dissipated as heat in the conductors. The values of the current and voltage at any moment of the discharge are also calculable from the equations given. With the capacities and inductances usually available, the oscillations of the current are of such high frequency as to constitute a jig. It is not, however, in the least necessary that an oscillatory current should have a high frequency. Mr J. H. Gray,* for example, has shown how to produce oscillatory currents having a period of about a second and a half by using large inductance and capacity in the circuit.

Let us now consider shortly Lord Kelvin's solution of the problem, which he has kindly given me permission to reproduce:—

"Let C be the capacity of the electrified body in electromagnetic measure, Q_0 its charge, and let it be suddenly connected to the earth by a wire of resistance R, the inductance of which is L. At a given moment t, the charge of the conductor is Q, and its potential $\gamma = \frac{Q}{C}$, which gives the equation for the potential—

$$\gamma = \frac{Q}{C} = L\frac{di}{dt} + Ri \qquad (1)$$

from Ohm's law and the definition of inductance.

* *Electrical Engineer*, August 19, 1892.

"Now the current $i=\frac{dQ}{dt}$, hence equation (1) may be written in the form

$$\frac{d^2Q}{dt^2}+\frac{R}{L}\frac{dQ}{dt}+\frac{Q}{CL}=0 \quad - \quad - \quad (2)$$

The general integral of this equation is—

$$Q=Ae^{\rho t}+A'e^{\rho' t},$$

ρ and ρ' being the roots of the quadratic equation,

$$\rho^2+\frac{R}{L}\rho+\frac{1}{CL}=0, \qquad (3)$$

which gives

$$\rho=-\frac{R}{2L}\pm\sqrt{\frac{R^2}{4L^2}-\frac{1}{CL}},$$

Reducing the fractions under the radical sign to a common denominator, we see that, according as $L \lessgtr \frac{R^2C}{4}$, these roots are real or imaginary. The constants A and A′ are determined by the condition that for $t=0$, we have $Q=Q_o$, *i.e.*, the initial charge, and $i=0$, which gives $0=A\rho+A'\rho'$.

"If the roots of the equation (3) are real, then representing by α the radical $\sqrt{\frac{R^2}{4L^2}-\frac{1}{CL}}$,

$$\left.\begin{aligned} Q&=Q_oe^{-\frac{Rt}{2L}}\left[\left(\tfrac{1}{2}+\frac{R}{4\alpha L}\right)e^{\alpha t}+\left(\tfrac{1}{2}-\frac{R}{4\alpha L}\right)e^{-\alpha t}\right], \\ i&=\frac{Q_o}{2\alpha LC}e^{-\frac{Rt}{2L}}(e^{\alpha t}-e^{-\alpha t}) \quad - \quad - \end{aligned}\right\}(4)$$

When the roots are imaginary we may still take the integral in the same form and replace the constants by their imaginary values. Putting $\alpha'=\sqrt{\frac{1}{CL}-\frac{R^2}{4L^2}}$, we get then

$$\left.\begin{aligned} Q&=Q_oe^{-\frac{Rt}{2L}}\left[\cos\alpha' t+\frac{R}{2\alpha' L}\sin\alpha' t\right] \quad - \quad - \\ i&=\frac{Q_o}{\alpha' LC}e^{-\frac{Rt}{2L}}\sin\alpha' t \quad - \end{aligned}\right\}(5)$$

"The nature of the discharge is very different according as the solution is given by equations (4) or equations (5). In the first case the discharge is continuous. The current begins by being zero, passes through a maximum, and then decreases to zero. The maximum is at the time determined by the condition $\frac{di}{dt}=0$, or

$$\left(\frac{R}{2L}-a\right)e^{2at}=\frac{R}{2L}+a,$$

which gives

$$t=\frac{1}{2\left(\frac{R^2}{4L^2}-\frac{1}{CL}\right)^{\frac{1}{2}}}l.\frac{\frac{R}{2L}+\sqrt{\frac{R^2}{4L^2}-\frac{1}{CL}}}{\frac{R}{2L}-\sqrt{\frac{R^2}{4L^2}-\frac{1}{CL}}}$$

"In the second case the values of Q and of i are given by the periodic functions; the conductor takes alternately charges in contrary directions, and the wire is the seat of alternating currents.

"*Period and Frequency.*—The times of maxima and minima of charge correspond to $i=0$, that is to say, to

$$\sin a't=0, \qquad \text{or} \qquad a't=n\pi.$$

"The oscillations of the discharge are therefore regular, and the value of the complete period T is

$$T=\frac{2\pi}{a'}=\frac{2\pi}{\sqrt{\frac{1}{CL}-\frac{R^2}{4L^2}}}$$

Hence the frequency $n=\frac{1}{T}=\frac{\sqrt{\frac{1}{CL}-\frac{R^2}{4L^2}}}{2\pi}$, or if R be small in comparison to L and C, $n=\frac{1}{2\pi\sqrt{CL}}$.

"*Damping.*—The values of the alternate maxima of charge are

$$+Q_0,$$
$$-Q_0e^{-\frac{R\pi}{2La'}}$$
$$+Q_0e^{-\frac{2R\pi}{2La'}}$$
$$-Q_0e^{-\frac{3R\pi}{2La'}}, \text{ \&c.}$$

They decrease, therefore, like terms of a geometrical progression, the ratio of which is $e^{-\frac{R\pi}{2La'}}$.

"The maxima of the intensity of the current in the two directions correspond to $\frac{di}{dt}=0$, which gives

$$\tan a't=\frac{2La'}{R},$$

or

$$\sin a't=\pm a'\sqrt{CL};$$

They are still equidistant, separated by a semi-period $\frac{\pi}{a'}=\frac{T}{2}$, and succeed the periods at which the current is zero, by the time θ, defined by the smallest angle which satisfies the condition

$$\sin a'\theta=a'\sqrt{CL}.$$

"The values of the maxima of current are successively

$$I_1=+\frac{Q_o}{\sqrt{CL}}e^{-\frac{R\theta}{2L}},$$

$$I_2=-\frac{Q}{\sqrt{CL}}e^{-\frac{R}{2L}\left(\theta+\frac{\pi}{a'}\right)}=-I_1e^{-\frac{R\pi}{2a'L}},$$

and so on; they also decrease in geometrical progression.

"Disregarding the sign, the total quantity of electricity put in motion is

$$Q_o\left(I+2e^{-\frac{R\pi}{2La'}}+2e^{-\frac{2R\pi}{2La}}+\ .\ .\ .\right)=Q_o\frac{I+e^{-\frac{R\pi}{2La'}}}{I-e^{-\frac{R\pi}{2La'}}}.$$

"This total mass is the greater the nearer the quantity $e^{-\frac{R\pi}{2La'}}$, is to unity, that is, the greater the factor $\frac{2La'}{R}$, or the greater is the ratio $\frac{L}{CR^2}$."

Hence inductance prolongs while capacity and resistance shorten the jig.

Transformers.—The equations for the mutual interaction of two circuits may be obtained in a somewhat similar way. For each circuit we have in addition to the terms of the form $L\frac{di}{dt}$, Ri, E; a term $M\frac{di^1}{dt}$ representing the action of the other circuit on the one considered. We have thus two simultaneous equations to solve. For transformers of very few turns, in which the electric induction (as opposed to the magnetic induction) is negligible, such equations give an approximate solution even with jigs; but if there is a considerable number of turns in either coil of the transformer, as in receiving jiggers, the dielectric induction between the layers must be taken into account as well as the magnetic induction between the coils.

From a theoretical point of view Lord Rayleigh has indicated the importance of electrostatic capacity in the construction of transformers for high frequency currents, and Mr Tesla has recognised its influence in practice, the form of the secondary winding of his H.F. transformer (Fig. 16, p. 26) being very like Fig. 109 (below).

In 1898, knowing the author's experience in electrical research, Mr Marconi asked him to assist in the development of his system. The problem which in particular required solution at that time was the production of a receiving transformer. Attempts had previously been made to construct one, but without any lasting success; and the direct connection of the coherer to the earth and aerial wires was still the method in use.

After a few weeks' work transformers were constructed which not only obviated the difficulties caused by the gradual collection of atmospheric electricity on the aerial, but also greatly increased the sensibility of the receiver, and later on made tuning possible.

A short research on the part played by the capacity of the adjacent layers of insulated wire which formed the windings of the transformer, showed that the condenser formed by two short layers of very fine wire, wound one on top of the other, had sufficient capacity to carry the whole jig current received without the aid of a conducting connection between them. It was clear, therefore, that capacity must play an extremely important rôle in the internal action of the transformer, that, in fact, the electric lines of force of the primary must be considered as well as the magnetic lines.

By consideration of the electric and magnetic induction between the primary and secondary coils of a small transformer, consisting of only two layers of equal length, it may be shown that during a complete alternation of the current the electric lines of force of the primary, and con-

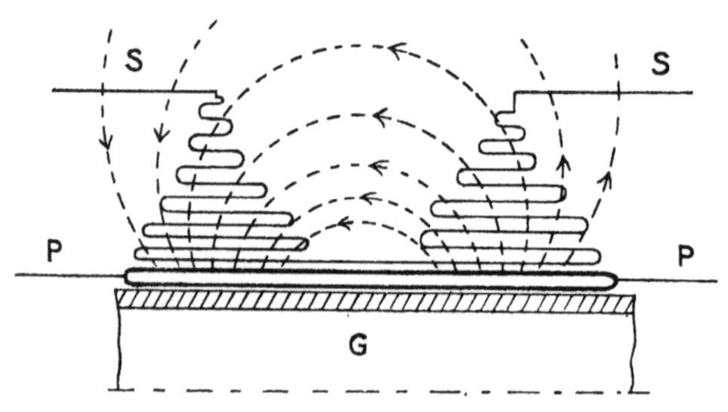

FIG. 109.—HALF SECTION OF RECEIVING TRANSFORMER (JIGGER), DESIGNED TO UTILISE ELECTROSTATIC INDUCTION.

P, Primary in two Parallel Layers; *S*, Secondary in Layers of Diminishing Length; *G*, Glass Tube.

[*N.B.*—Layers of wire are shown as full lines, and not as rows of dots, which would actually represent them.] Dotted lines show direction of electric force at time of maximum.

sequently the current induced by them, are mainly in opposition to the E.M.F. resulting from the varying magnetic induction due to the primary current. It was clearly advantageous to give the secondary some other shape so that the electrically and magnetically induced currents in the secondary might be brought nearly into coincidence.

One way in which this was accomplished is shown in Fig. 109. The dotted lines indicate the direction of the electric force between the ends of the primary at times when it is a maximum, and indicate how the form of the secondary takes advantage of it.

In another form the secondary is wound on one half of the primary only, the other half being under the first and

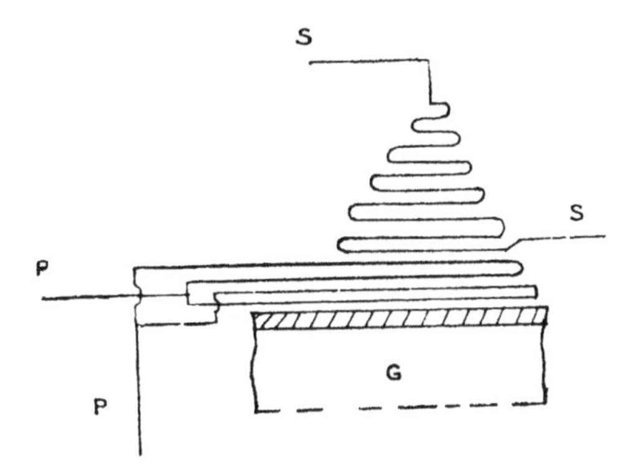

FIG. 110.—HALF SECTION OF RECEIVING TRANSFORMER (JIGGER).

P, *P*, Primary in Two Double Layers, Earth Ends next Glass Tube; *S*, *S*, Secondary; *G*, Glass Tube.

thus screened from the secondary. The explanation is the same in both cases.

The author then suggested the form shown in Fig. 111,

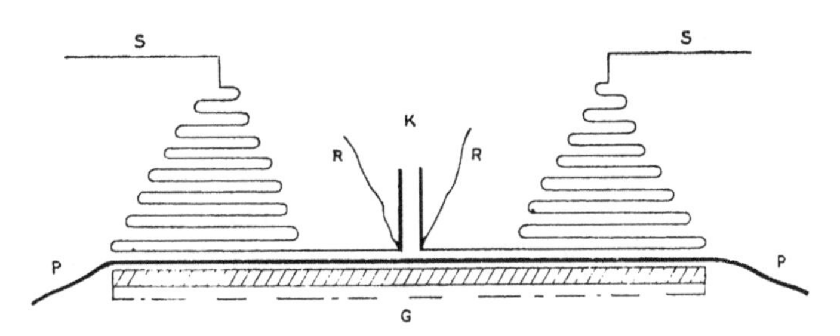

FIG. 111.—HALF SECTION OF RECEIVING TRANSFORMER WITH SPLIT SECONDARY.

P, *P*, Primary; *S*, *S*, Secondary; *K*, Condenser; *R*, *R*, Leads to Relay Circuit; *G*, Glass Tube.

the object of placing the condenser in the middle of the secondary instead of at one end, as had been previously the custom, was to permit of both ends of the secondary being connected directly to the coherer terminals so that the coherer should be at the position where maximum voltage is developed. This arrangement appears to have been adhered to by the Marconi Company. Measurements, in which the author was not directly concerned, of the lengths of the secondaries of successful jiggers showed that in order to obtain the best results it was necessary to give the length of the secondary a definite relation to the length of the aerial, thus making revibration possible.

In other forms of winding, as shown in Fig. 27, the electric and magnetic inductions assist one another, and of course the length of the secondary is such that the receiver circuit has the same frequency of vibration as the aerial.

A successful receiving jigger is thus a resonating transformer designed so that the electric induction assists, or at least is not in opposition to, the current produced by change of magnetic induction.

Fig. 27, page 43, shows a direct coupled primary and secondary in which the central part of the single layer forms the primary, and is also a part of the secondary. In this case the capacity between primary and secondary is practically negligible, but there is a direct conducting connection instead.

The construction of transmitting transformers is somewhat simpler, the currents dealt with being much larger, and high efficiency not so essential. Those used by most companies are simply modifications of Mr Tesla's original high frequency resonating transformer. It is needless to go into their slight variations in detail. The most interesting is Mr Tesla's most recent pattern, the design of which is shown in Chapter XVIII. It will be noticed that the end of the secondary which attains high potentials is at the centre of the disc, and is thus separated by the whole radius of the disc from the earthed end. Thus the difference of potential between adjacent turns of the secondary is comparatively small, and very high voltages may be obtained without difficulties due to faulty insulation or to capacity.

Rate of Sparking—Number of Jigs per Second.—Ordinary methods of transmission involve the production of a number of sparks per second. Each spark represents a train of gradually decreasing oscillations of high frequency—what we have called a jig, in fact—and is associated with a definite amount of energy, which is the same for every jig. If, therefore, the apparent or integral current due to the passage of a succession of jigs through the measuring instrument be increased by increasing the number of jigs per second without altering the total energy of each, we should find that the current registered is proportional to the square root of the number per second.

If we double the number of jigs per second we double the watts, or activity; we do not, however, double the current, as shown by an instrument indicating R.M.S. current, but merely increase it in proportion to the square root of the number per second, for the activity is proportional not to the current directly, but to its square.

That this has been proved experimentally to be the case is shown by the following extract from an exceedingly important paper by Messrs Duddell and Taylor, communicated to the Institution of Electrical Engineers in 1905:—

"*Effect of the Rate of Sparking.*—With the strongly coupled method of transmitting the ammeter indicated, a R.M.S. current in the transmitter aerial of 2.31 amperes for a rate of sparking of 42 per second; an increase in the speed of the interrupter to give 67 sparks per second, keeping the spark length constant, increased the R.M.S. current to 2.83 amperes. Thus, although the rate of sparking was changed in the ratio of 1 : 1.60, the current in the aerial only altered in the ratio of 1 : 1.22. Using the weak method of coupling, a similar result was obtained. In this case an increase in the spark rate in the ratio 1 : 1.67 caused the current to increase in the ratio 1 : 1.30. It is to be noted that in both cases the R.M.S. current in the aerial varied approximately proportional to the square root of the rate of sparking. This also applies to the current in the receiver aerial. If we now wish to consider how the R.M.S. current in the aerial during the train of oscillations produced by each interruption of the current of the induction coil is affected by the rate of sparking, we must multiply the observed currents by the square root of the ratio of the whole time between one spark and the next to the time the rate of oscillation lasts. Assuming that the time each individual train of oscillations lasts is unaffected by the spark rate, we see that the R.M.S. current in the transmitter or receiver aerial produced by each interruption of the coil current, was practically independent of the spark rate."

Lieutenant Tissot has most kindly permitted me to quote the following from a paper communicated by him to the Institution of Electrical Engineers in January 1906:—

"My experimental observations also showed that if the number of interruptions is made to vary from n to n' per second, the effective value of the current received by the aerial varies in the ratio of $\sqrt{n}$ to $\sqrt{n'}$. This result, which appears in accord with the observations of Duddell and Taylor, is readily capable of interpretation by reasoning as follows. Let

I denote the reading on a hot-wire ammeter put in series at the bottom of the transmitting wire.
A = the amplitude of the current in the aerial.
T = the period.
y = the decrement of the oscillation.
n = number of interruptions, that is of wave trains per second.

"By a simple integration we then obtain for the fundamental wave—

$$I^2 = \frac{nA^2}{4y} \cdot \frac{4\pi^2}{4\pi^2 + y^2} \cdot T.$$

"Since the factor $\frac{4\pi^2}{4\pi^2 + y^2}$ is almost equal to unity, it is clear that the energy transmitted by a single wave train, which is proportional to A^2, is given by a relation of the form $W_e = KA^2 = K'\frac{I^2}{n}$ (K and K′ being constants). In the same manner the energy received for a single wave train is easy to compute, if one assumes that the detector completely absorbs the energy. If an effective current i is obtained with n interruptions, that is, with n trains per second, then the equation $W=\frac{pi^2}{n}$ expresses the energy received for a single wave train, p being the resistance of the measuring instrument (in this case the bolometer), when the number of interruptions varies from n to n', i varies from i to i', and $\frac{i'^2}{i^2}=\frac{n'}{n}$, whence $\frac{i'}{i}=\frac{\sqrt{n'}}{\sqrt{n}}$."

CHAPTER XVII.

ON THEORIES OF TRANSMISSION.

IN Chapter I. some suggestions have been made as to the way in which the electrical energy sent out by the transmitter may reach the receiver. Opinions still differ considerably as to the actual mechanism of the process, as for instance, in regard to the part played by the earth, and the importance of free as opposed to conducted radiation; and a large number of facts as to variations in the possible distance of transmission on account of intervening hills and atmospheric conditions have yet to be allotted their proper places in the general scheme.

The remarkable fulfilment of a mathematical prophecy which we shall consider in the next chapter is a unique case in regard to the electrical transmission of energy for telegraphy or other purposes, and presents less mathematical difficulty than the more ordinary systems. It is these latter, however, that we must deal with at present, and in doing so it will be well to commence with a survey of the electrical properties of the system of conductors and dielectrics with which we have to deal.

The Atmosphere.—This is a thin spherical shell composed of a rather complex mixture of gases, bounded internally by the surface of the earth or sea, and externally by what is probably a nearly perfect vacuum. Well known facts in regard to meteorites and other cosmical phenomena show that the thickness of the shell is not much over 100 miles; it is therefore thin in proportion to its diameter, which is approximately 8,000 miles, and thin also in proportion to the distances of over 2,000 miles which are common in wireless telegraphy. At the earth's surface the air pressure is about 15 lbs. on the square inch, and the density is about .08 lb. per cubic foot; at the outer surface of the shell, say 100 miles above the earth, both pressure and density are zero. The pressure and density at any intermediate point are calculable by the well-known barometric formula.

The electrical properties of air at various pressures are well known. At 15 lbs. on the square inch it is, except

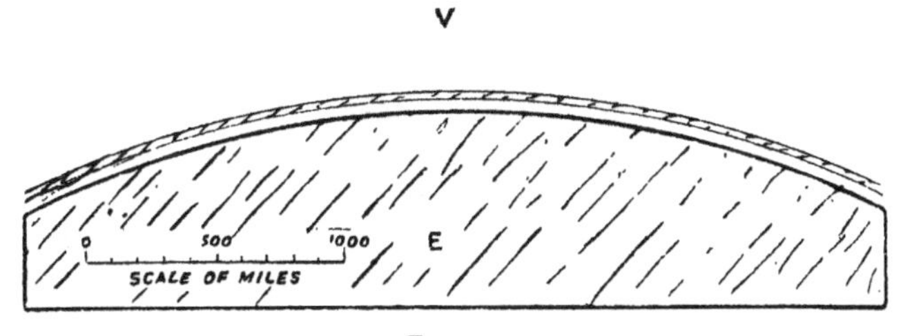

FIG. 112.

A portion of the Earth and Atmosphere, to scale. *E*, the Earth; *V*, Outer Space. The shaded parts are conductors, the strip between them being the dielectric of Wireless Telegraphy.

for extremely high voltages, a nearly perfect dielectric or non-conductor—that is to say, lines of electric force can travel through it without appreciable loss by dissipation of energy. At a low pressure such as exists at about 35 miles above the earth, it becomes suddenly conductive, and at the pressure which must exist a few miles higher up its conductivity, for high frequency currents without electrodes, is as great as that of a 25 per cent. solution of sulphuric acid, so great, in fact, that a layer of half an inch thick forms a complete screen through which electric waves do not pass.* At lower pressures, about 80 miles up, it becomes such a non-conductor as we know it in "hard"

* J. J. Thomson, "Recent Researches in Electricity."

X-ray tubes, and, farther up still, as perfect a dielectric as is known. There is a critical value of the pressure, which occurs at about 35 miles, below which height the air is practically a non-conductor, and above which its conductivity rapidly attains that of the best electrolytes. The *non-conducting* shell of air which surrounds the earth is therefore only about 35 miles thick—less than one hundredth part of its radius—and is bounded below by earth, which has a resistance of about 6,600 ohms per c.c., or sea, with 373 ohms per c.c.,* and above by a layer having a resistance of not more than 10 ohms per c.c. The upper shell then conducts at least 660 times as well as damp earth, and nearly 40 times as well as the sea.

This intermediate non-conducting shell of air is the dielectric with which we have to deal in wireless telegraphy. Though in general a good insulator, its properties in this respect are by no means constant, its ionisation, and consequently its conductivity, being subject to both regular and irregular changes. The most important of the former is the diurnal variation corresponding to the daily variation of atmospheric electrification. The maxima at the earth's surface occur between 8 and 10 A.M., and between 10 P.M. and 1 A.M., the minima being at 2 P.M. and 4 A.M. It has been found that even air in closed vessels, as well as in the open, is subject to these variations.† In both cases the effect is most probably due, directly or indirectly, to streams of electrified particles, ejected by the sun, impinging on the upper atmosphere; the daily variation being caused by the rotation of the earth presenting different parts in turn to the rays.

The maxima and minima just given are those observed near the earth's surface. Higher up, the variation, as far as is known, has only one maximum and one minimum daily, the former about noon and the latter at 5 A.M. The times of maximum conductivity are those at which there is

* Brylinski, *Soc. Internat. Elect. Bulletin*, 6, p. 255, June 1906.

† A. Wood, *Nature*, 73, p. 583, April 19, 1906.

greatest dissipation of energy in the transmission of electric force through the dielectric. This occurs in the air not in contact with the earth every day about noon. The time of minimum conductivity occurs about 5 A.M., and this is also the time, as might be expected, at which wireless signals travel farthest.

It may seem curious that increase of conductivity of the dielectric causes increase of dissipation of energy, and therefore less efficient transmission; the difficulty will be at once removed, however, when it is remembered that in all modes of transmission of electrical energy it is necessary to have good insulation, and that any leakage in the insulator causes waste, and prevents a proportion of the energy reaching the desired point. The discoverer of this difference between day and night was Mr Marconi, who found, while on a voyage to America, that signals which failed entirely during the day at distances of more than 700 miles were easily obtained up to 1,500 miles during the night. The time at which the greatest change occurred was about 7 A.M. at the transmitting station, which, as the experiments were made about the 1st of March, was just after sunrise there. He therefore put forward the theory that the weakening must be due to the action of daylight on the transmitting aerial itself. Professor J. J. Thomson has, however, shown that the waste of energy due to ionisation of the air should be greatest where the electrical energy is most concentrated, *i.e.*, near the transmitting aerial, and that it increases with the wave-length of the transmitted jig.

The phenomenon is therefore explainable as simply on this as on Marconi's hypothesis, and the fact that it is only observable in transmission to long distances favours the supposition that the loss does not only take place at the aerial but is to some extent distributed throughout the whole course.

Captain Jackson's observations on similar alterations in the possible distance of signalling, due not to day and night but to other changes in atmospheric conditions, are very interesting. He has found that the maximum signalling distance may be reduced from 65 to 22 miles in apparently fine weather, though as a rule, such changes are the accompaniment of a falling barometer and the approach of stormy conditions.

It was also found that during a scirocco wind, which is damp and laden with dust from the African coast, the distance was very considerably reduced.

In both cases a change in the properties of the atmosphere is evidently the cause of the variation, and apparently the change is one which causes increased conductivity and therefore greater dissipation of energy of the waves which are propagated through the air.

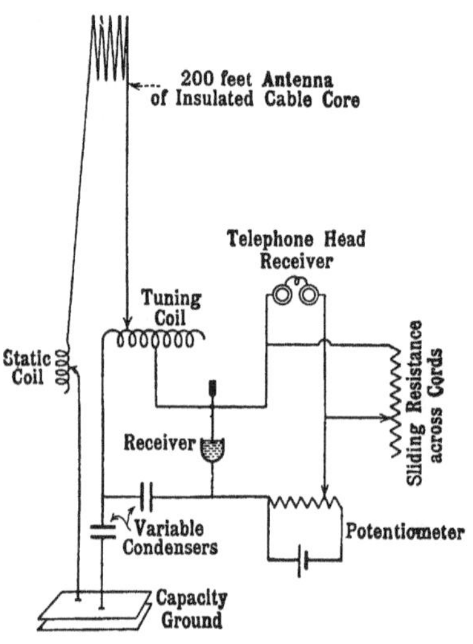

Fig. 113.—Captain Wildman's Method of Testing the Intensity of Signals.

Captain Wildman, of the United States Army, has found that in damp and stormy weather the transmission of signals becomes more difficult. His observations were made across Norton Sound, in Alaska, between stations at Safety and St Michael, the distance being 107 miles. By shunting the receiving telephone (as shown in Fig. 113) the strength of the signals could be reduced to a point at which they were just audible. The conductivity of the shunt required indicates the strength of current in receiving aerial, as it shows how much the current may be cut down without causing signals to become inaudible.

Early in March 1906 Captain Wildman kindly sent me

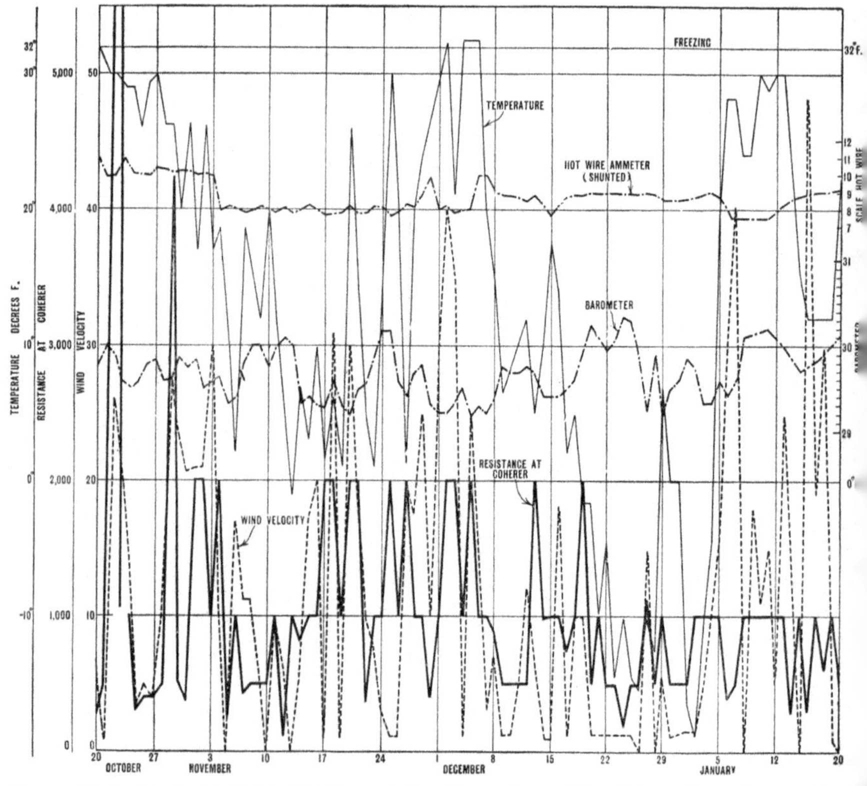

FIG. 114.—DIAGRAM ILLUSTRATING OBSERVATIONS MADE BY CAPTAIN WILDMAN (SEE HIS "NOTES") ACROSS NORTON SOUND, IN ALASK

To face page 227.]

the diagram (Fig. 114) and notes given below. The obvious conclusion which may be drawn from these results is that stormy damp weather reduces the strength of signals. This coincides with Captain Jackson's observations, and as the receivers used in the two cases were entirely different it is safe to locate the cause of the variation outside the instrument room, *i.e.*, either on the aerials, or, more probably, in the atmosphere.

NOTES BY CAPTAIN WILDMAN.—The thin line representing temperatures can be easily traced, the horizontal being marked in the extreme left hand column, directly underneath the lines 32.

The upper dot and dash line is a record of the hot-wire ammeter at the sending station. It is merely a relative measure, and does not read directly in amperes. A 2-ampere Jones and Phillipps hot-wire ammeter, with a large shunt uncalibrated for highly oscillating currents, was used. In practice, the inductance in the closed oscillating and open radiating circuits are moved until the maximum reading on the hot-wire ammeter is obtained.

The lower dot and dash line represents barometer readings, and its scale is between 29 and 31, marked at the extreme left.

The broken line represents wind velocity in miles per hour, with its scale nearest the margin of the chart.

The thick line represents resistance on coherer obtained by putting non-inductive resistance across the telephone cords.

This gives the relative measure of loudness in the receiver.

The points on the chart marked with a * are practically the only points where discrepancies occur in the parallelism of the wind and resistance curves.

Those on the 25th of November and the 13th of December are easily explained as errors in the reading of the operator, or much noise in the room, making acute hearing an impossibility.

That on 5th January I am unable to explain, unless the day were very, very dry, and the wind velocity made no difference, taken in connection with the fact that the hot-wire ammeter reading was greater than it had been for many days. It is also possible that the operator, by practice, had been able to cut down the sound in the telephone receiver, to a greater extent than when the observations were first commenced. On the whole, the two lines parallel each other so well, that there is no doubt in my mind that there is some connection between the wind velocity and the travelling, or about to travel, electromagnetic wave.

Great variations in the strength of the received current have also been observed at the Transatlantic stations of the Fessenden system. The curve (Fig. 98, p. 182) shows the enormous variations which may occur in the course of a few days, due to causes apparently beyond the control of the operators; *e.g.*, to atmospheric changes.

The Earth.—Let us now consider the electrical properties of the other factor concerned in transmission, viz., the earth. This body is, roughly, a sphere of about 4,000 miles radius. Its properties as a single electrical conductor will be considered in the next chapter, in the meantime we shall confine ourselves to the conductivity of its upper layers, since these only are concerned in ordinary wireless telegraphy, because high frequency currents do not penetrate far into a conductor.

M. Brylinski, in an exhaustive paper on the resistance of conductors to variable currents,* discusses thoroughly the resistance of the earth and sea to currents of low and high frequencies, whether uniformly alternating or damped. Assuming the value 6,600 ohms per c.c., based on experiments, for the resistance of damp earth, and unit permeability, M. Brylinski shows that if the frequency of alternation of a current be 10 millions per second, which corresponds roughly to those employed in wireless transmission, practically the whole current will be confined to a layer not more than 15 metres deep. If the current be damped, the depth is decreased, and therefore the effective resistance is increased.

The depth of the useful layer in salt water is about one quarter of that in earth, *i.e.*, about 4 metres; but as the conductivity of salt water is about seventeen times as great as that of soil, the effective resistance is much less.

If the soil contain iron or other magnetic material, the depth of penetration is enormously decreased.

These facts probably account in great part for differ-

---

* *Soc. Int. Elect. Bulletin*, 6, pp. 255-300, June 1906

ences in the possible distance of transmission over hills of various materials which were observed by Captain Jackson in a most interesting series of experiments, which we shall consider later.

Stated in words, M. Brylinski's principal result is as follows:—In spite of the infinite depth of the conductor, the resistance of a strip of finite width has a finite and calculable value which increases with the specific resistance, with the frequency, with the permeability, and with the damping. He has also calculated a similar formula for the inductance of the useful layer.

Thus we see that if a jig of given frequency and damping is conducted along the earth's surface, the resistance of the layer which carries the current is increased if the specific resistance or permeability is increased. Now let us examine Captain Jackson's observations, which are the result of a large number of experiments made between vessels fitted with what is practically his own system of wireless telegraphy. In each case the maximum distance of transmission over sea was known, so that the reduction due to any intervening land could be found by direct experiment. In all cases a certain proportion of the distance, generally much the greater part, was sea, a fact which must be kept in mind in what follows. Experiments such as these, showing actual dissipation of energy *en route*, in addition to the reduction due to the increasing diameter of the wave, assist greatly in making clear the method of propagation of the waves. The chief results may be submarised thus:—

If the possible distance over sea be taken as 100, the distance when sandstone or shale intervened was 72, with hard limestone intervening 58, and with limestone containing iron ores, 32.

Now, recollecting that sandstone has a much higher resistance than sea water, and limestone than sandstone, and that limestone containing iron ores has both high resistance and permeability, we are at once driven to the conclusion, by comparison of these figures with M. Brylinski's formula, that the distance to which wireless transmission is possible is governed to a large extent by the resistivity and permeability of the useful upper layer of the earth or sea. Since, therefore, the nature of the earth's surface determines the distance to which signals can be transmitted, it is clear that whatever be the distribution and shape of the lines of force in the atmosphere, transmission depends directly on the action of the earth as a conductor—at least in an earthed system such as was used by Captain Jackson.

Having established so much, let us consider the course of the lines of force which accompany the current. In

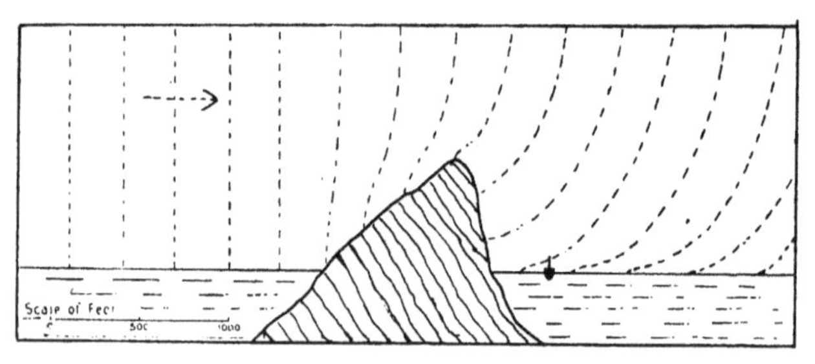

FIG. 115.—DIAGRAM EXPLANATORY OF ONE OF CAPTAIN JACKSON'S EXPERIMENTS ON SCREENING. (Roughly to Scale.)

general a line of electric force is at right angles to a conductor where it meets it; in our case, therefore, they are approximately vertical to the surface of the earth or sea. If they are, as we suppose, travelling with their ends continuously attached to the earth, they will be at right angles to the surface whether this surface be level or not. Also if sharp bends occur in the surface, the lines will also be bent at a short distance above the surface.

One of Captain Jackson's observations appears to bear directly on this latter question. It is a case in which signals suddenly failed when the motion of the receiving vessel brought a high promontary of land between her and the transmitter. I have made a sketch, roughly to scale, from Captain Jackson's description of this experiment showing the position of the ship and the cliff, and I have put in approximately the moving lines of force. It will be seen that the mast and aerial wire on the ship are at such an angle to the lines of force that even if the resistance of the land had not previously reduced the current, no signals could be received. The aerial practically lies parallel to the surface of the earth (the face of the cliff), with the insulated end pointing towards the direction from which the waves come. This position is, as is well known, an extremely disadvantageous one for the reception of signals. That the bending of the lines and not their absence is, in part at least, the cause of this apparent screening, is rendered still more probable by the observation that though no signals could be received close to the land, in one instance, they could be obtained at a greater distance from the transmitter by moving the ship away from the land in a straight line. Our explanation is that the lines have approximately regained their vertical direction.

If, in the first experiment, Captain Jackson had inclined the aerial away from the cliff at a considerable angle he would probably have received signals.

It is also true that these sudden changes of direction of motion cause some dissipation of energy owing to the induction of what are practically eddy currents in the conducting surface, and that any bend in a conductor means an increase of inductance. Recent experiments by Mr De Forest and Mr Marconi, described in Chapter XV., with horizontal wires laid on the ground or at only a short distance above it, entirely confirm the theory that in these systems of telegraphy the lines of force are attached to the earth, and are guided by it.

Leaving for a time the consideration of the earth currents, and of the field of electric force immediately above the earth's surface, let us see what must occur in the intermediate and upper regions of the atmosphere.

Firstly we must notice that by an application of Lord

Kelvin's method of "Electrical Images," which appears to be perfectly legitimate in the circumstances, it has been shown that the form of the lines of force in the case of a vertical aerial conductor with spark-gap at the bottom is the same as would be produced if a second aerial equally, but oppositely charged, and of the same dimensions, were run vertically downwards from the spark-gap, and the earth removed altogether. This statement, of course, only holds exactly if the surface of the earth be a plane conducting surface, hills and variations in the nature of the surface will make certain modifications and corrections necessary. The lines of force may thus be taken from Hertz's diagrams by drawing a horizontal line through the middle point of the oscillator to represent the earth's surface, and obliterating the part of the diagram below it: their form is shown in Figs. 10-13, page 18.

As these hemispherical waves attain larger and larger diameter their wave-length remains constant, hence the width of the strip on which the base of the wave stands is constant. Thus at the earth's surface the only variation is in the circumference of the strip, and as this is inversely proportional to the distance from the transmitter we find that the electric force and the displacement at any point of the surface are inversely proportional to the distance from the transmitter. The energy per unit volume, which is proportional to the product of the force into the displacement, is therefore proportional to the square of the distance.

This holds only for the immediate neighbourhood of the surface, if the wave be spherical; higher up the energy becomes less and less dense, and is zero in the vertical line above the aerial. The propagation of this type of electrical disturbance has been investigated by Heaviside and by Blondel. Heaviside has also suggested that an upper conducting layer of the atmosphere might prevent the radiation of electrical energy into space, and so modify the spherical form of the waves, rendering them ultimately cylindrical at a considerable distance from the origin.* In a paper communicated to the Institution of Electrical Engineers, in December 1905, the author independently restated this proposition, and adduced some arguments in its favour, including the experimental results of Professor J. J. Thomson mentioned above, and the observations of auroræ recently made by Danish meteorologists, which prove that auroræ are limited to the layer of the atmosphere between six and sixty miles above the earth's surface, and therefore that this layer is conductive; since then a further proof, which will be considered in the next chapter, has been given of the part played by the upper atmosphere, confirming by direct electrical experiment the conclusions deduced in the earlier part of this chapter from physical considerations.

Messrs Duddell and Taylor† found from experiments made between a station at Howth and H.M.T.S. "Monarch" in the Irish Channel, that the received current received varies at distances beyond ten or fifteen miles, in almost exact simple inverse proportion to the distance between the stations. The result admits of no question, the experiments having been carried out most carefully and with wonderfully concordant results. They have been fully confirmed by Lieutenant Tissot.‡ The measurements were made with Mr Duddell's thermo-ammeter and thermo-galvanometer, the former in the transmitting aerial and the latter in the receiver. The greatest distance reached was Holyhead, sixty miles from the station at Howth. The results of these experiments are of so great importance that with the kind permission of the authors I reproduce the diagrams which give the details. There are many interesting points to be noted in connection with them, explanations of some of which we may be able to give from consideration of our deductions from the work of M. Brylinski and Captain Jackson.

* "Encyclopædia Britannica," art. Telegraphy, vol. 33, p. 215.

† *Journal of the Institution of Electrical Engineers*, No. 174, vol. 35, July 1905.

‡ *Ibid.* No. 177, vol. 36, April 1906.

Admitting then, as is generally done by those best qualified to judge, that the application of the method of electrical images, described above, to the problem of determining the form of the lines of force during their outward motion, is legitimate, we see that waves must on the earth's surface consist of a series of concentric circles whose centre

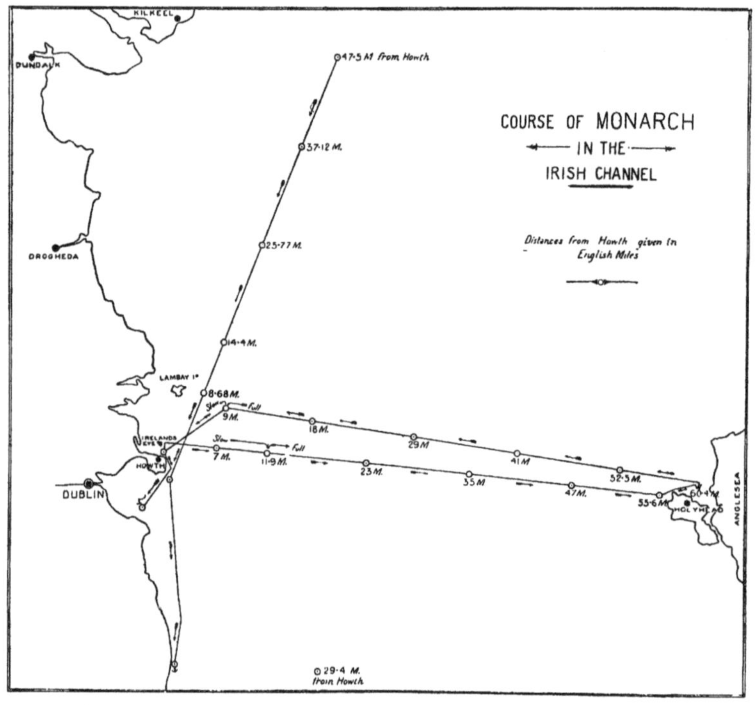

FIG. 116.—DUDDELL AND TAYLOR'S SKETCH CHART, HOWTH TO HOLYHEAD.

is the transmitter. The width of the circular strip containing the jig will be approximately constant, and will depend on the number of waves in the jig and on the wave-length. In the transmission of a jig to distances greater than (wave-length) × (number of waves in the jig), the last wave will have left the transmitter before the first has arrived at the receiver. Hence in this case the receiver has no influence on the transmitter. This is generally the case

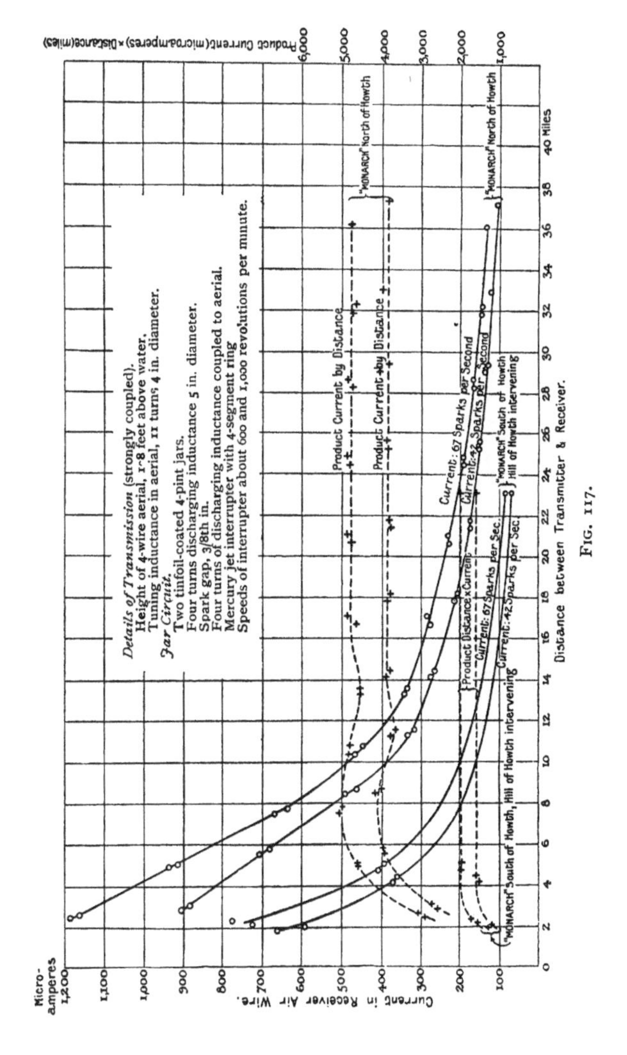

FIG. 117.

at distances above a few miles. Fig. 120 shows roughly the form of the wave fronts of the jig in two positions—(1) when

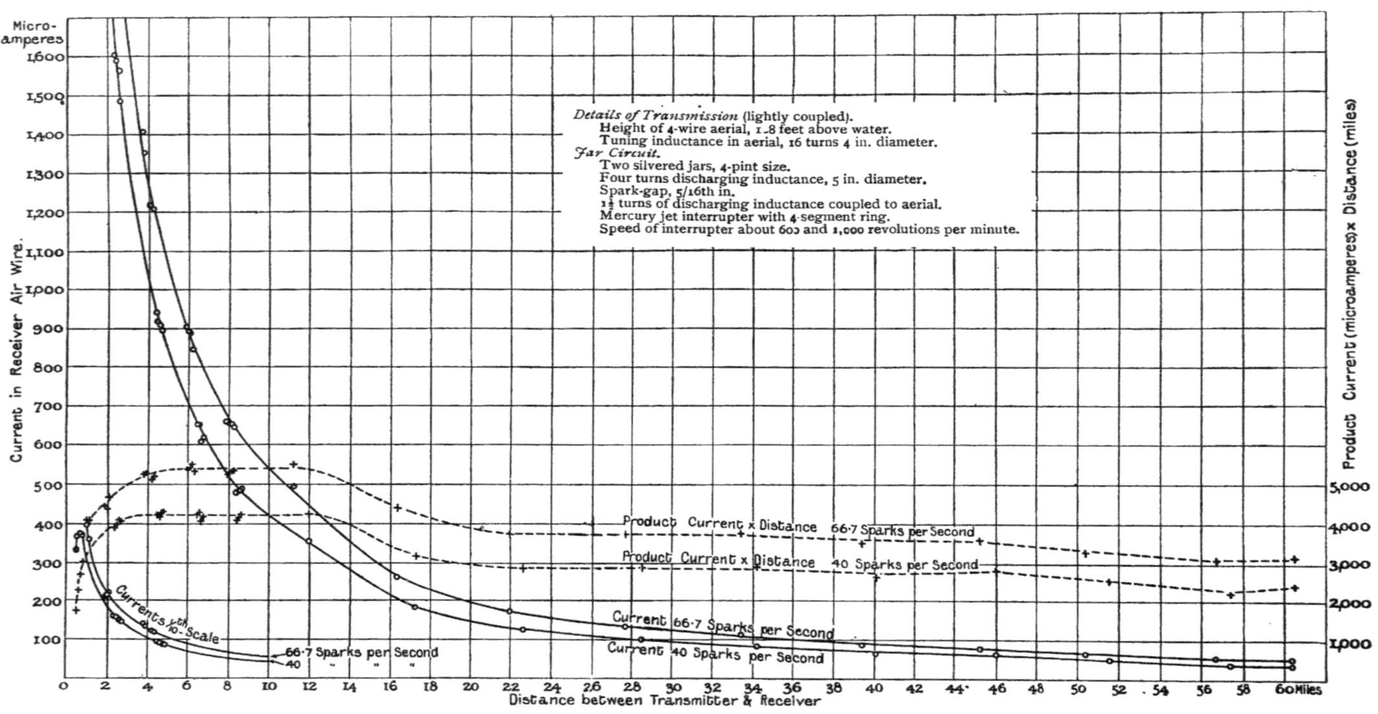

FIG. 118.—"Monarch" returning from Holyhead to Howth.

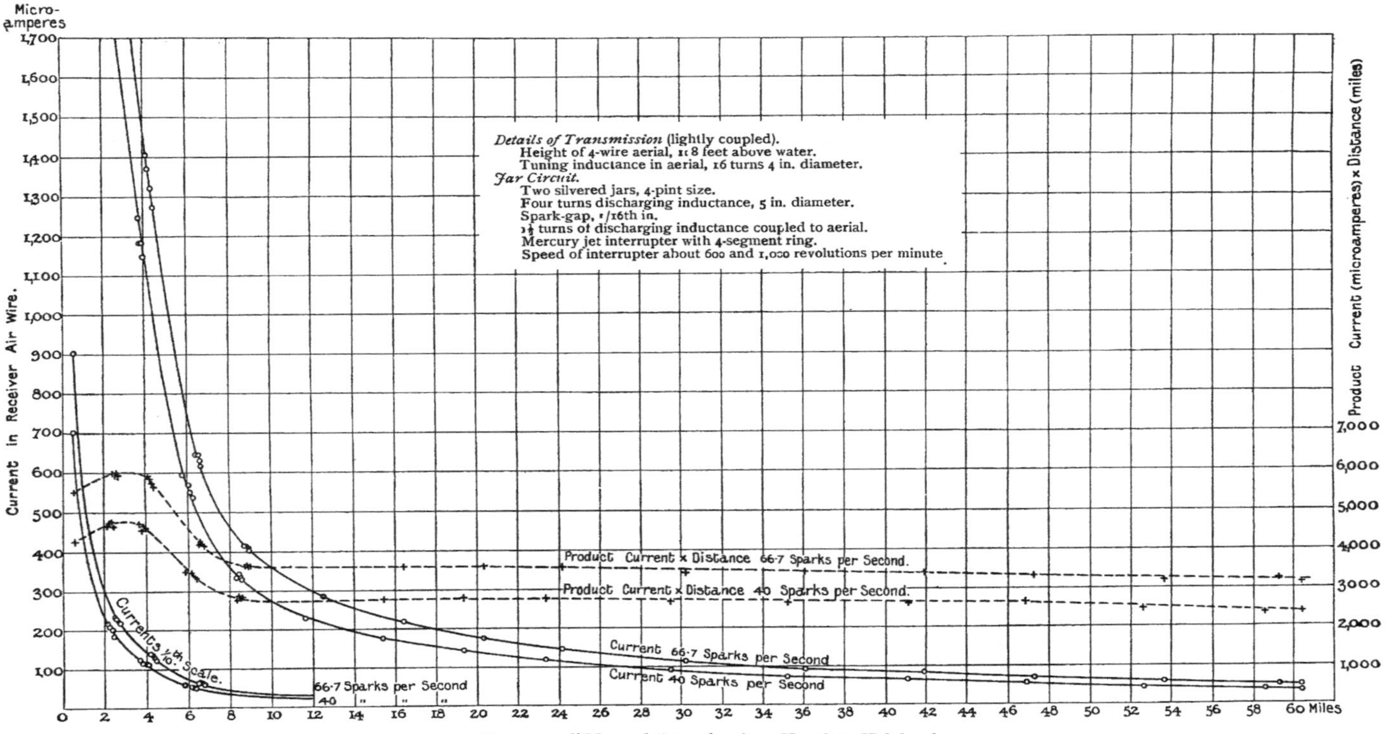

FIG. 119.—"Monarch" crossing from Howth to Holyhead.

just clear of the transmitter, and (2) when it has commenced to affect the receiver. From this diagram, and from the fact that the jig current is confined to a thin layer of the earth's surface, we see that it is the resistance of a strip stretching from one station to the other, and more particularly the resistance of a V-shaped slab near each station that influences the propagation of the jig, and therefore the received current. Let us first take the case of two stations both of which are situated on an interrupted plain or on the

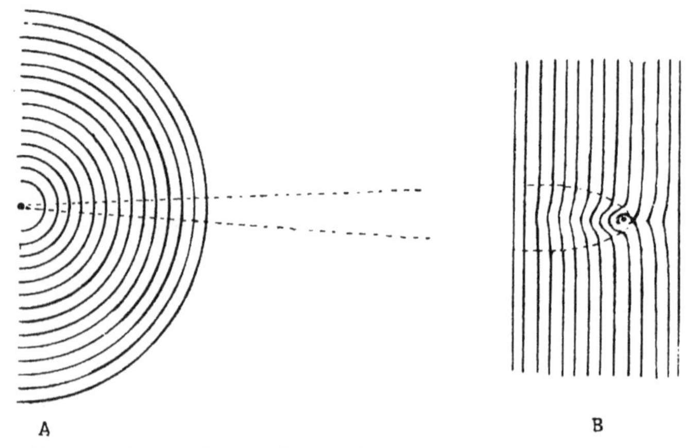

FIG. 120.—DIAGRAM SHOWING PLAN OF PROPAGATION OF JIGS ALONG EARTH'S SURFACE.

A, Jig leaving Transmitter; *B*, Jig affecting Receiver. The dotted lines enclose the "useful strip" of earth whose resistance directly effects transmission.

sea. The jig from the transmitter travels outwards unmodified except by its increasing circumference and by the slight losses due to frictional dissipation of energy caused by the resistance of the earth and the imperfect insulation of the atmosphere. Neglecting the frictional losses for the present, we see that theory, confirmed by Duddell and Taylor's experiments, gives a variation in the electric force, and therefore in the received current, which is inversely proportional to the distance from the transmitter. We shall call this the law of divergence, as it shows the loss of energy due to purely geometrical considerations. It may be expressed symbolically as

$$c=\frac{C}{d}$$

where c is the current in the receiving aerial, d the distance of transmission, and C a constant.

If we take the curvature of the earth into account, and proceed to trace the variation of the wave front at greater distances than those of Duddell's experiments, which only extended to sixty miles, we find from simple mathematical, or rather geographical, considerations that the radius of the circle in which the wave front cuts the globe is directly proportional to the sine of the angle at the earth's centre subtended by the arc of the surface between the transmitter

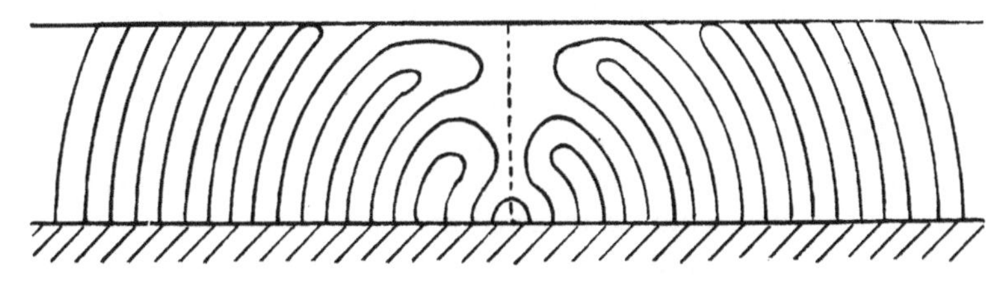

FIG. 121.

and the wave front. This agrees with Duddell's law, because for small distances $\sin\theta$ is nearly equal to θ. For greater distances, however, the divergence becomes less and less, and at distances beyond 6,000 miles, *i.e.*, beyond $\theta=\frac{\pi}{2}$, the waves converge, and the current should increase with distance. This is true on any theory whatever, as the wave front is becoming a smaller and smaller ring, continually converging from all sides towards the antipodes of the transmitting station.

The equation for the received current is thus of the form :—

$$c=\frac{C}{\sin\theta}; \qquad \text{(A)}$$

where c is the received current, θ the angle defined above, and C a constant.

Fig. 121 gives a rough representation of the lines of electric force as they spread from the transmitter outwards.

The above law of variation of the received current with distance is obviously not complete. We have so far supposed that the lowest layer of the atmosphere is a perfect dielectric, and that the earth and upper atmosphere are perfect conductors, *i.e.*, that no dissipation of energy is involved, the decrease of current being simply due to increase in area of the wave front.

We must now take account of frictional losses, in order to complete the law as far as possible; though with the data at present available only a rough approximation can be attempted. It will be possible to amplify the equation when further measurements have been made, and to determine more exactly the functions and constants involved.

The first indication of a true frictional dissipation of energy in transmission was found in Marconi's discovery that messages could be received at much greater distances during the night than by day. There cannot be much difference between the divergence variations by night and day, as the earth does not change its shape; this phenomenon has therefore to be explained by some difference in the electrical properties of the transmitting media. Now the conductivity of the earth cannot be seriously altered by the sun's rays; the atmosphere, however, may be much affected. Cosmical science has made many striking advances lately, and it is now practically certain that streams of electrified particles, resembling the cathode rays, are ejected from the sun with a very high velocity. These penetrate the earth's atmosphere on the day side, electrifying it, and at the same time ionising it, thus rendering it more conductive. During the night equilibrium becomes slowly restored, and towards five o'clock in the morning a very marked minimum of atmospheric electrification has long been known to exist. The time at which signals go farthest thus coincides with a minimum conductivity of the lower layers of the atmosphere, while greater conductivity of the lower layers during the daytime coincides with greater difficulty in transmission.

Now an electric wave must have either an all-dielectric medium or a dielectric bounded by conductors; it cannot penetrate far through a conductor. Thus if all the atmosphere were conductive the wave would travel only a very short distance before being dissipated by resistance, and transmission to a great distance would be impossible. The conditions would be similar to an attempt to transmit current along a concentric cable with bad insulation between the conductors. The good insulation is just as essential as the conductors. We may take it, then, that during the night the lowest ten miles of the atmosphere is a good dielectric, while during the day it becomes slightly conductive.

To obtain a rough estimate of the amount of this dissipation, we may take Mr Marconi's statement, that 500 miles during the day is roughly equivalent to 1,000 by night. The variation is not so noticeable at short distances, being in this case so very much less important than the divergence, though it is distinctly traceable in Duddell and Taylor's results.

Let us assume that the law of divergence is that given above—Equation (A)—*i.e.*, that the undissipated current would vary inversely as the sine of the angle between the places, or approximately, for the distances under consideration, inversely as the distance along the earth's surface.

We then have Duddell's law—

$$c \times d = \text{constant},$$

where c is current received, and d is distance between the stations.

Now assume as a first approximation that there is no dissipation at night; then the current lost, by divergence only, between 500 and 1,000 miles at night is all spent by dissipation during the day before 500 miles is reached. Again, the divergence alone between 500 and 1,000 miles would reduce the current to one-half, since 1,000 is twice 500. Therefore the current at 500 miles during the night is twice the current which will just work the receiver. The current lost by dissipation during the day throughout 500 miles is therefore equal to that which will just work the receiver. We have thus obtained an approximate value of the dissipation from Mr Marconi's experiments. If we call the minimum working receiver current of Mr Marconi's receiver M, then the current lost through dissipation is roughly $\frac{M}{500}$ per mile.

This dissipation is no doubt a function of the distance, the form of which I hope to be able to determine later. It is also, of course, a function of the state of the atmosphere, as regards electrification, and varies from a minimum in the very early morning to a maximum in the evening. In the meantime we may take $\frac{M}{500}$ as a rough approximation to its average mid-day value, under the circumstances considered.

The approximate equation $c = \frac{C}{\sin \theta}$ for propagation of signals then becomes—

$$c = \frac{C}{\sin \theta} - \frac{M}{500} d, \qquad \text{(B)}$$

where—

c = received current,
C = a constant,
M = minimum current which will actuate receiver,
d = miles distance,

and—

θ = angle subtended at centre of earth by arc d.

Now since $\frac{d}{3500} = \theta$, approximately, since the earth's radius is 3,500 nautical miles, $\therefore d = 3{,}500\,\theta$.

To determine C from (A) we notice that $c = C$, if $\sin\theta = 1$, *i.e.*, if $\theta = \frac{\pi}{2}$. C is thus the value of c at 5,400 nautical miles, or an earth quadrant from the transmitter during the night, *i.e.*, when *no* dissipation. We may state it as $C = kM$ if we like to take M as a unit, *i.e.*, the undissipated current at 5,400 nauts. is k times that necessary to work the receiver.

The equation then becomes—

$$c = M\left(\frac{k}{\sin\theta} - 7\theta\right) \qquad \text{(C)}$$

To find the limit of distance with any given power we must have $c = M$, *i.e.*, the current at the receiver is just the smallest possible working current.

The equation (C) then becomes:

$$1 = \frac{k}{\sin\theta} - 7\theta, \qquad \text{(D)}$$

or—

$$k - \sin\theta\,(7\theta + 1) = 0 \qquad \text{(D}^1\text{)}$$

The number seven in this equation is only a rough approximation, suitable to the case under consideration. It should properly be replaced by some function of θ.

Let us now consider Duddell and Taylor's results for shorter distances in the light of this explanation.

Remarking on the slight decrease in the product $c \times d$ at long distances, they suggest reflection from the Hill of Howth as a cause, mentioning that another possible, but apparently less probable, explanation is that the cross-channel runs were made in damp, misty daylight, while the up-and-down channel course was made in the night and early morning with a clear frosty atmosphere. To me this latter suggestion is far more probable than the first. We know that such a difference exists from Marconi's results. It is worth while to attempt to derive it from Messrs Duddell and Taylor's measurements. We may take from their curves the following data, comparing the day and the night experiments.

(1.) *Night.*

"MONARCH" COMING SOUTH TOWARDS HOWTH.

Distance off Howth	18	...	36	Miles.
$c \times d$	3,850	...	3,820	Micro-amperes × Miles.

Thus at 18 miles the current was 214 mA., and at 36 miles 106.1 mA.

Now, by divergence alone the current would at 36 miles have been one-half that at 18 miles, *i.e.*, 107 mA. The dissipation loss is thus 0.9 mA. in 18 miles, or roughly 0.05 mA. per mile on 214 mA.

(2.) *Day.*

In a similar way we may obtain by averaging curves, representing the same quantities on the voyage to Holyhead and back during the day, the value 7.2 mA. lost between 30 and 60 miles off Howth. This amounts roughly to 0.24 mA. per mile on a maximum current of 120 mA. Reducing the result given in (1) to suit this maximum current, we see that the dissipation by night is to that by day in the ratio of about .025 to .24, *i.e.*, the dissipation during the day is approximately ten times as much as at night.

Messrs Duddell and Taylor attempted to measure the day and night variation between Howth and Kingstown, but did not find as much as 1 per cent. difference. This was natural, as these points are only about six miles apart, hence the total dissipation was very small in both cases. The difference of the values for night and day which I have deduced above would barely amount to 1 per cent. of the total in five miles.

Comparing the value of the dissipation determined from Marconi's experiments with that obtained from Duddell and Taylor's, we find that they are of the same order of magnitude. Nearer than this we cannot at present go, since the minimum working current of Marconi's long-distance receiver is not known.

Effects of Unsymmetrical Distribution of Resistance in the Neighbourhood of the Receiving Station. —Messrs Duddell and Taylor attribute the unexpectedly large currents indicated by the humps in the curves in Figs. 118 and 119, to effects of reflection from the mainland of Howth Head or the islands. When we consider, however, how comparatively slight is the slope of even a "precipitous" headland we see that waves travelling in straight lines would, on striking the slope, be shot up skywards and could by no means reach the receiving station on the earth. Captain Jackson's suggestion made in the discussion on the paper, that the humps might be due to the interference of the two jigs of different wave-length sent out simultaneously by every transmitter, is rendered improbable as an explanation in the present case by the fact that the distance over which the hump stretches is so different in the two voyages of the "Monarch."

I would suggest that the following explanation is free from either objection, and explains the facts more simply.

The Howth receiving station was situated on land near the sea (see sketch map), at a place where the coast line runs approximately east and west. There was therefore land on the east and south sides of the station. Fig. 122 shows how this configuration of the land and sea would affect, by the difference of resistance, the current transmitted. The upper shaded elliptical surface shows the part of the useful strip of the earth's surface near the receiver at the end of the voyage from Holyhead. In this case it is almost entirely sea. The same is true at the commencement of the outward voyage towards Ireland's Eye.

The other shaded portion of the diagram shows that at seven miles out the strip now includes, near the receiver, a large piece of land of high resistance. There is not much further variation in the land included, as the voyage is continued; there is also but little variation in electrical conditions as shown by the curve of Fig. 119.

On the return voyage a more northerly course was taken, and Howth was "opened up" at a much greater distance; thus the amount of land included in the course of the current near the receiver was the same at twenty miles

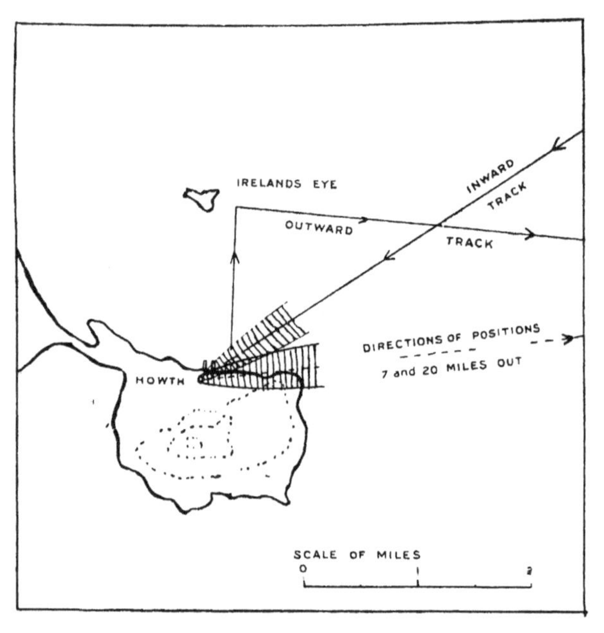

FIG. 122.—DISTRIBUTION OF RESISTANCE NEAR RECEIVER. USEFUL PORTIONS OF WAVE TRAINS ARRIVING AT HOWTH FROM VARIOUS DIRECTIONS.

off Howth on the return journey as it had been at seven miles out on the outward trip. The inclusion of a greater length of sea in the useful strip makes but little difference in the total resistance, since the resistance of sea water is so low; the resistances in the two cases are therefore practically equal.

Now the products (received current) (distance), whose

values indicate the losses of energy from causes other than simple geometrical divergence, are, when the "Monarch" was at the positions just mentioned, exactly equal; the "other cause" is therefore almost certainly the *resistance* of the useful strip. The coincidence seems too clear to be the result of chance, and a minute study of the curves and map at smaller and greater distances of transmission, gives nothing but confirmation of the view that under similar atmospheric and instrumental conditions, the resistance of the useful strip is the most important factor, next to geometrical divergence itself, in determining the current received. A comparison, for instance, of the values of the product $c \times d$ for positions north and south of Howth, with that for the intermediate case just cited shows a similar relation. Thus, north of Howth, where the useful strip near the receiver is all sea, $c \times d = 480$; east of Howth, where nearly half the strip is land, $c \times d = 360$; and south of Howth, with useful strip all land, $c \times d = 200$. These numbers indicate an almost exact proportionality between the received current and the conductance of the useful strip near the receiver. Many facts well known to wireless telegraphists, as, for instance, that it is better to place a station on a flat shore near the sea than on a hill farther inland, afford independent evidence of the correctness of this theory.

CHAPTER XVIII.

WORLD-WAVE TELEGRAPHY.

MR TESLA in the year 1905 patented a system of wireless telegraphy and distribution of electrical energy which utilises the whole world as a conductor by the creation of stationary electric waves on it. In 1893 he suggested the possibility of doing so, in 1899 he discovered experimentally that such waves are produced by thunderstorms, and in 1906 he states that he has so far perfected his apparatus as to be able to produce them artificially.

What had previously been deduced theoretically in regard to such waves may be given in the words of one of our best known physicists, the late Professor G. F. Fitzgerald. In 1893 Professor Fitzgerald wrote * as follows :—

"The period of oscillation of a simple sphere of the size of the earth, supposed charged with opposite charges of electricity at its ends, would be about $\frac{1}{17}$ of a second; but the hypothesis that the earth is a conducting body surrounded by a non-conductor is not in accordance with fact. Probably the upper regions of our atmosphere are fairly good conductors. . . . If we assume the height of the region of the aurora (*i.e.*, the upper conducting layer) to be 60 miles, or 100 kilometres, we get a period of oscillation of 0.1 sec. Assuming it to be six miles (or 10 km.), the period becomes 0.3 sec. . . . Dr Lodge has already looked for evidence . . . on the assumption that the period would be $\frac{1}{17}$ sec.; but with a negative result."

---

* *Nature*, September 28, 1893.

Sir Oliver Lodge's experiments were made at a period when the resources of the wireless telegraph receiver were unknown. Some years later Mr Tesla, having developed a sensitive wireless receiver capable of being tuned to jigs of any frequency, erected an experimental station at Colorado Springs, and spent many months in observing the electrical disturbances produced in the earth by the extremely violent thunderstorms which are common in that district.

Through the courtesy of Mr T. C. Martin, Editor of the *Electrical World and Engineer*, New York, I am enabled to make the following quotations from Mr Tesla's only publications on the subject. In 1893 Mr Tesla said: *—

"Assume that a source of alternating currents be connected, as in Fig. 123, with one of its terminals to earth (conveniently to the water mains), and with the other to a body of large surface P. When the electric oscillation is set up there will be a movement of electricity in and out of P, and alternating currents will pass through the earth, converging to, or diverging from, the point *c* where the ground connection is made. In this manner neighbouring points on the earth's surface within a certain radius will be disturbed. But the disturbance will diminish with the distance, and the distance at which the effect will still be perceptible will depend on the quantity of electricity set in motion. Since the body P is insulated, in order to displace a considerable quantity the potential of the source must be excessive, since there would be limitations as to the surface of P. The conditions might be adjusted so that the generator or source S will set up the same electrical movement as though its circuit were closed. Thus it is certainly practicable to impress an electric vibration at least of a certain low period upon the earth by means of proper machinery. At what distance such a vibration might be made perceptible can only be conjectured. I have on another occasion considered the question how the earth might behave to electric disturbances. There is no doubt that, since in such an experiment the electrical density at

* "Inventions, &c., of Nikola Tesla," p. 348, T. Commerford Martin (New York, 1893).

the surface could be but extremely small considering the size of the earth, the air would not act as a very disturbing factor, and there would not be much energy lost through the action of the air, which would be the case if the density were great. Theoretically, then, it could not require a great amount of energy to produce a disturbance perceptible at great distance, or even all over the surface of the globe. Now, it is quite certain that at any point within a certain radius of the source S, a properly adjusted self-induction and capacity device can be set in action by resonance. But not only can this be done, but another source S_1, Fig. 123, similar to S, or any number of such sources, can be set to work in synchronism with the latter, and the vibration thus intensified and spread over a large area, or a flow of electricity produced to or from the source S_1 if the same be of opposite phase to the source S. I think that beyond doubt it is possible to operate electrical devices

FIG. 123.—TESLA'S SUGGESTED WIRELESS TELEGRAPHY.

in the city through the ground or pipe system by resonance from an electrical oscillator located at a central point. But the practical solution of this problem would be of incomparably smaller benefit to man than the realisation of the scheme of transmitting intelligence, or perhaps power, to any distance through the earth or environing medium. If this is at all possible, distance does not mean anything. Proper apparatus must first be produced, by means of which the problem can be attacked, and I have devoted much thought to this subject. I am firmly convinced that it can be done, and hope that we shall live to see it done."

In March 1905, Mr Tesla, describing his Colorado experiments of 1899, mentions that one evening a violent thunderstorm had gathered in the mountains and drifted eastwards over the observing station. When it had passed, he continues :—

FIG. 124.—MR TESLA'S ELECTRICAL OBSERVATORY AT COLORADO SPRINGS.

"The recording apparatus being properly adjusted, its indications became fainter and fainter with the increasing distance of the storm, until they ceased altogether. I was watching with eager expectation. Surely enough, in a little while the indications began again, grew stronger and stronger, and, after passing through a *maximum*, gradually decreased and ceased once more. Many times, in regularly recurring intervals, the same actions were repeated until the storm, which, as evident from simple computations, was moving with nearly constant speed, had retreated to a distance of about 300 kilometres. Nor did these strange actions stop then, but continued to manifest themselves with undiminished force. Subsequently similar observations were also made by my assistant, Mr Fritz Lowenstein, and shortly afterward several admirable opportunities presented themselves which brought out still more forcibly and unmistakably the true nature of the phenomenon. No doubt whatever remained; I was observing stationary waves.

"As the source of the disturbances moved away the receiving circuit came successively upon their nodes and loops. Impossible as it seemed, this planet, despite its vast extent, behaved like a conductor of limited dimensions." *

The electric observatory is shown in Fig. 124.

The actual results of this investigation are described in the British patent No. 8,200 of 1905, from which I quote the following:—

"It has long since been known that electric currents may be propagated through the earth, and this knowledge has been utilised in many ways in the transmission of signals and the operation of a variety of receiving devices, remote from the source of energy, mainly with the object of dispensing with a return conducting wire.

"It is also known that electrical disturbances may be transmitted through portions of the earth by grounding only one of the poles of the source, and this fact I have made use of in systems, which I have devised for the purposes of transmitting through the natural media intelli-

* *Electrical World and Engineer*, New York, March 5, 1904.

gible signals or power, and which are now familiar. But all experiments and observations heretofore made have tended to confirm the opinion held by the majority of scientific men, that the earth, owing to its immense extent, although possessing conducting properties, does not behave in the manner of a conductor of limited dimensions with respect to the disturbances produced, but, on the contrary, much like a vast reservoir or ocean which, while it may be locally disturbed by a commotion of some kind, remains unresponsive and quiescent in a large part or as a whole.

"Still another fact, now of common knowledge, is that when electrical waves or oscillations are impressed upon such a conducting path as a metallic wire, reflection takes place under certain conditions from the ends of the wire, and, in consequence of the interference of the impressed and reflected oscillations, the phenomenon of 'stationary waves,' with maxima and minima indefinite, fixed positions, is produced. In any case the existence of these waves indicates that some of the outgoing waves have reached the boundaries of the conducting path and have been reflected from the same.

"Now I have discovered that notwithstanding its vast dimensions, and contrary to all observations heretofore made, the terrestrial globe may, in a large part or as a whole, behave towards disturbances impressed upon it in the same manner as a conductor of limited size, this fact being demonstrated by novel phenomena which I shall hereinafter describe.

"In the course of certain investigations which I carried on for the purpose of studying the effects of lightning discharges upon the electrical condition of the earth, I observed that sensitive receiving instruments, arranged so as to be capable of responding to electrical disturbances created by the discharges, at times failed to respond when they should have done so, and upon inquiring into the causes of this unexpected behaviour, I discovered it to be due to the character of the electrical waves, which were produced in the earth by the lightning discharges and which had nodal regions following at definite distances, the shifting source of the disturbances. From data obtained in a large number of observations of the maxima and minima of these waves, I found their length to vary, approximately from 25 to 70 kilometres, and these results and certain theoretical deductions led me to the conclusion that waves of this kind may be propagated in all directions over the globe, and that they may be of still more widely differing lengths, the extreme limits being imposed by the physical dimensions and properties of the earth.

"Recognising in the existence of these waves an unmistakable evidence that the disturbances created had been conducted from their origin to the most remote portions of the globe and had been thence reflected, I conceived the idea of producing such waves in the earth by artificial means, with the object of utilising them for many useful purposes for which they are or might be found applicable.

"This problem was rendered extremely difficult owing to the immense dimensions of the earth and, consequently, enormous movement of electricity, or rate at which electrical energy had to be delivered, in order to approximate, even in a remote degree, movements or rates which were manifestly attained in the displays of electrical forces in nature, and which seemed at first unrealisable by any human agencies. But by gradual and continuous improvements of a generator of electrical oscillations, which I have described in the Specifications of my United States Patents, Nos. 645,576 and 649,621, and in the Specification of my British Patent, No. 24,421 of 1897, I finally succeeded in reaching electrical movements, or rates of delivery of electrical energy, not only approximating but, as shown in many comparative tests and measurements, actually surpassing those of lightning discharges, and by means of this apparatus I have found it possible to reproduce whenever desired phenomena in the earth the same as or similar to those due to such discharges.

"With the knowledge of the phenomena discovered by me, and the means at command for accomplishing these results, I am enabled not only to carry out many operations by the use of known instruments, but also to offer a solution for many important problems involving the operation or control of remote devices which, for want of this knowledge and in the absence of these means, have heretofore been entirely impossible.

"For example, by the use of such a generator of stationary waves and receiving apparatus, properly placed and adjusted

FIG. 125.—MR TESLA'S TRANSMITTING STATION ON LONG ISLAND.

in any other locality, however remote, it is practicable to transmit intelligible signals ; or to control or actuate at will any or all of such apparatus for many other important and valuable purposes, as for indicating wherever desired the correct time of an observatory ; or for ascertaining the relative position of a body or distance of the same with reference to a given point ; or for determining the course of a moving object, such as a vessel at sea, the distance traversed by the same or its speed ; or for producing many other useful effects at a distance dependent on the intensity, wave-length, direction or velocity of movement, or other feature or property of disturbances of this character.

"I shall typically illustrate the manner of applying my discovery by describing one of the specific uses of the same, namely, the transmission of intelligible signals or messages between distant points, and with this object reference is now made to the accompanying drawings, in which—

"Fig. 126 represents diagrammatically the generator which produces stationary waves in the earth,

"Fig. 127 an apparatus, situated in a remote locality, recording the effects of these waves, and

"Fig. 128 the usual arrangement of the circuits of my receiving transformer.

"In Fig. 126 A designates a primary coil forming part of a transformer, and consisting generally of a few turns of a stout cable of inappreciable resistance, the ends of which are connected to the terminals of a source of powerful electrical oscillations, diagramically represented by G. This source is usually a condenser charged to a high potential, and discharged in rapid succession through the primary, as in a type of transformer invented by me and now well known, having been described in my patents on apparatus of this kind, of which it will be sufficient to mention my British Patent, No. 20,981 of 1896. But when it is desired to produce stationary waves of great lengths, an alternating dynamo of suitable construction may be used to energise the primary A.

"C is a spirally wound secondary coil within the primary, having the end nearer the latter connected to the ground E, and the other to an elevated terminal D. The physical constants of coil C, determining its period of vibration, are so chosen and adjusted that the secondary system E C D is in the closest possible resonance with the oscillations impressed upon it by the primary A. It is, moreover, of the greatest importance in order to still further enhance the rise of pressure and to increase the electrical movement in the secondary system, that its resistance be as small as practicable, and its self-induction as large as possible under the conditions imposed. The ground should be made with great care, with the object of reducing its resistance.

"Instead of being directly grounded, as indicated, the coil C may be joined, in series or otherwise, to the primary A, in which case the latter will be connected to the plate E. But be it that none, or a part, or all of the primary or exciting turns are included in the coil C, the total length of the conductor from the ground plate E to the elevated terminal should be equal to one-quarter of the wave-length of the electrical disturbance in the system E C D, or else equal to that length multiplied by an odd number. This relation being observed, the terminal D will be made to coincide with the points of maximum pressure in the secondary or excited circuit, and the greatest flow of electricity will take place in the same. In order to magnify the electrical movement in the secondary as much as possible, it is essential that its inductive connection with the primary A should not be very intimate, as in ordinary transformers, but loose, so as to permit free oscillation. That is to say, their mutual induction should be small. The spiral form of coil C secures this advantage, while the

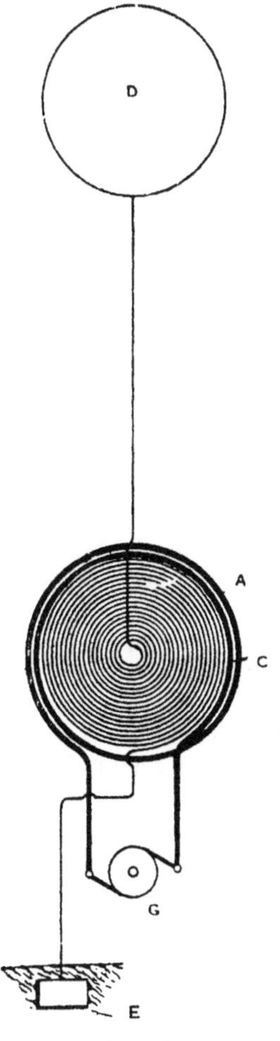

FIG. 126.

turns near the primary A are subjected to a strong inductive action and develop a high initial electromotive force.

"These adjustments and relations being carefully completed, and other constructive features indicated rigorously observed, the electrical movement produced in the secondary system by the inductive action of the primary A will be enormously magnified, the increase being directly proportionate to the inductance and frequency, and inversely to the resistance of the secondary system. I have found it practicable to produce in this manner an electrical movement thousands of times greater than the initial, that is the one impressed upon the secondary by the primary A, and I have thus reached activities or rates of flow of electrical energy in the system E C D, measured by many tens of thousands of horse-power. Such immense movements of electricity give rise to a variety of novel and striking phenomena, among which are those already described. The powerful electrical oscillations in the system E C D being communicated to the ground, cause corresponding vibrations to be propagated to distant parts of the globe, whence they are reflected, and by interference with the outgoing vibrations, produce stationary waves, the crests and hollows of which lie in parallel circles, relatively to which the ground plate E may be considered to be the pole. Stated otherwise, the terrestrial conductor is thrown into resonance with the oscillations impressed upon it just like a wire. More than this, a number of facts ascertained by me clearly show that the movement of electricity through it follows certain laws with nearly mathematical rigour. For the present it will be sufficient to state that the earth behaves like a perfectly smooth or polished conductor of inappreciable resistance, with capacity and self-induction uniformly distributed along the axis of symmetry of wave propagation and transmitting slow electrical oscillations without sensible distortion and attenuation. Besides the above, three requirements seem to be essential to the establishment of the resonating condition.

"1. The earth's diameter passing through the pole [*i.e.*, station] should be an odd multiple of the quarter wavelength, that is, of the ratio between the velocity of light and four times the frequency of the currents.

"2. It is necessary to employ oscillations, in which the

rate of radiation of energy into space in the form of Hertzian or electro-magnetic waves is very small. To give an idea I would say, that the frequency should be smaller than twenty thousand per second, though shorter waves might be practicable. The lowest frequency would appear to be six per second, in which case there will be but one node, at or near the ground plate, and, paradoxical as it may seem, the effect will increase with the distance, and will be greatest in a region diametrically opposite the transmitter. With oscillations still slower the earth, strictly speaking, will not resonate, but simply act as a capacity, and the variation of potential will be more or less uniform over its entire surface.

"3. The most essential requirement is, however, that irrespective of frequency, the wave or wave train should continue for a certain interval of time, which I have estimated to be not less than one-twelfth—or probably 0.08484—of a second, and which is taken in passing to and returning from the region diametrically opposite the pole, over the earth's surface, with a mean velocity of about 471,240 kilometres per second.

"The presence of the stationary waves may be detected in many ways. For instance, a circuit may be connected directly, or inductively, to the ground and to an elevated terminal, and tuned to respond more effectively to the oscillations. Another way is to connect a tuned circuit to the ground and to two points lying more or less in a meridian passing through the pole E, or generally stated, to any two points of a different potential.

"In Fig. 127 I have shown a device for detecting the presence of the waves, such as I have used in a novel method of magnifying feeble effects, which I have described in my United States Patents, Nos. 685,953 and 685,955, and my British Patent, No. 11,293 of 1901. It consists of a cylinder C^2 of insulating material which is moved at a uniform rate of speed, by clockwork or other suitable motive power, and is provided with two metal rings B, B^1, upon which bear brushes A and A^1, connected respectively, to the terminal plates P and P^1. From the rings B and B^1 extend narrow metallic segments S and S^1, which, by the rotation of the cylinder C^2 are brought alternately into contact with double brushes B and B^1, carried by, and in contact with, conducting holders H and H^1, supported in metallic bearings D and D^1, as shown. The latter are connected to the terminals T and T^1 of a condenser C, and it should be understood that they are capable of angular displacement, as ordinary brush supports. The object of using two brushes, as B and B^1, in each of the holders H and H^1, is to vary at will the duration of the electric contact of the plates P and P^1, with the terminals T and T^1, to which is connected a receiving circuit including a receiver R, and a device performing the duty of closing the receiving circuit at predetermined intervals of time, and discharging the stored energy through the receiver. In the present case this device consists of a cylinder D made partly of conducting and partly of insulating material E and

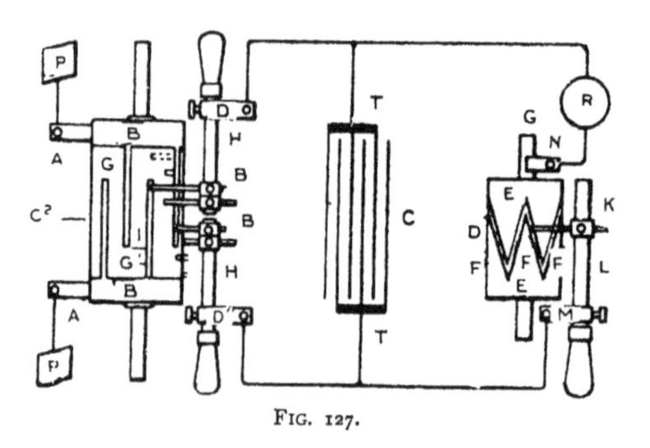

FIG. 127.

E^1, respectively, which is rotated at the desired rate of speed by any suitable means. The conducting part E is in good electrical connection with the shaft S, and is provided with tapering segments F, F, F, upon which slide a brush K supported on a conducting rod L, capable of longitudinal adjustment in a metallic support M. Another brush N is arranged to bear upon the shaft S, and it will be seen that whenever one of the segments F comes in contact with the brush K, the circuit including the receiver R is completed, and the condenser discharged through the same. By an adjustment of the speed of rotation to the cylinder D, and a displacement of the brush K along the cylinder the circuit may be made to open and close in as rapid succession, and remain open or closed during such intervals of time, as may be desired.

"The plates P and P^1, through which the electrical energy is conveyed to the brushes A and A^1, may be at a considerable distance from each other in the ground, or one in the ground and the other in the air, preferably at some height. If but one plate is connected to the earth and the other maintained at an elevation, the location of the apparatus must be determined with reference to the position of the stationary waves established by the generator, the effect evidently being greatest in a maximum, and zero in a nodal region. On the other hand, if both plates be connected to earth, the points of connection must be selected with reference to the difference of potential, which it is desired to secure, the strongest effect being, of course, obtained when the plates are at a distance equal to half the wave-length.

"In illustration of the operation of the system, let it be assumed that alternating electrical impulses from the generator are caused to produce stationary waves in the earth, as above described, and that the receiving apparatus is properly located with reference to the position of the nodal and ventral regions of the waves. The speed of rotation of the cylinder C^2 is varied until it is made to turn in synchronism with the alternate impulses of the generator, and the position of the brushes B and B^1 is adjusted by angular displacement, or otherwise, so that they are in contact with the segments S and S^1 during the periods when the impulses are at, or near, the maximum of their intensity. These requirements being fulfilled, electrical charges of the same sign will be conveyed to each of the terminals of the condenser, and with each fresh impulse it will be charged to a higher potential. The speed of rotation of the cylinder D being adjustable at will, the energy of any number of separate impulses may thus be accumulated in potential form and discharged through the receiver R upon the brush K coming into contact with one of the segments F. It will be understood that the capacity of the condenser should be such as to allow the storing of a much greater amount of energy than is required for the ordinary operation of the receiver. Since by this amount a relatively great amount of energy, and in suitable form, may be made available for the operation of a receiver, the latter need not be very sensitive. But, when the impulses are very weak, or when it is desired to operate a receiver very rapidly, any of the well-known sensitive devices, capable of responding to very feeble influences, may be used by the manner indicated or in other ways.

"Under the conditions described it is evident that during the continuance of the stationary waves, the receiver will be acted upon by current impulses more or less intense, according to its location with reference to the maxima and minima of said waves, but upon interrupting or reducing the flow of the current the stationary waves will disappear or diminish in intensity. Hence a great variety of effects may be produced in a receiver according to the mode in which the waves are controlled. It is practicable, however, to shift the nodal and ventral regions of the waves at will from the sending station, as by varying the length of the waves under observance of the above requirements. In this manner the regions of maximum and minimum effect may be made to coincide with any receiving station or stations. By impressing upon the earth two or more oscillations of different wave-length a resultant "stationary" wave may be made to travel slowly over the globe, and thus a great variety of useful effects may be produced. Evidently, the course of a vessel may be easily determined without the use of a compass, as by a circuit connected to the earth at two points, for the effect exerted upon the circuit will be greatest when the plates P, P^1 are lying on a meridian passing through ground plate E, and will be *nil* when the plates are located at a parallel circle. If the nodal and ventral regions are maintained in fixed positions the speed of a vessel carrying a receiving apparatus may be exactly computed from observations of the maxima and minima regions successively traversed. This will be understood when it is stated, that the projections of all the nodes and loops on the earth's diameter passing through the pole, or axis of symmetry of the wave-movement, are all equal. Hence in any region at the surface the wave-length can be ascertained from simple rules of geometry. Conversely, knowing the wave-length, the distance from the source can be readily calculated. In like ways the distance of one point from another, the latitude and longitude, &c., may be determined from the observation of such stationary waves. If several such generators of stationary waves—

preferably of different lengths—were installed in judiciously selected localities, the entire globe could be subdivided in definite zones of electric activity, and such and other important data could be at once obtained by simple calculation or readings from suitably graduated instruments.

"The specific plan of producing the stationary waves, herein described, might be departed from. For example, the circuit which impresses the powerful oscillations upon the earth might be connected to the latter at two points.

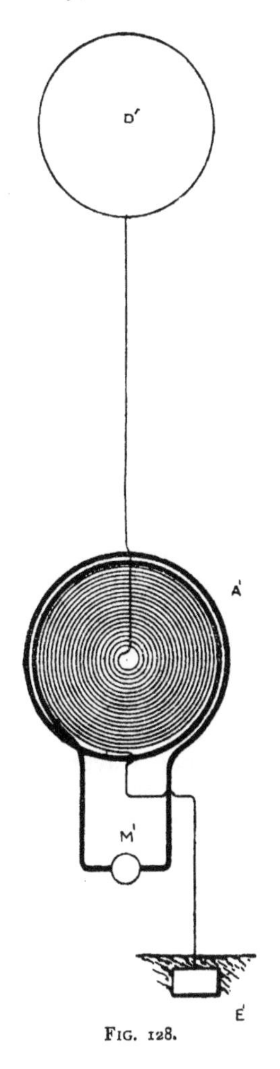

FIG. 128.

"In collecting the energy of these disturbances in any terrestrial region at a distance from their source, for any purpose, and, more especially, in appreciable amounts, the most economical results will be generally secured by the employment of my synchronised receiving transformer. This invention, forming part of my system of transmission of energy through the natural media, has been fully explained in the patents first cited here, but for the better understanding of the present description it is diagrammatically illustrated in Fig. 128. Its most essential part is a circuit E^1 C^1 D^1, which is connected, arranged, and adjusted similarly to the transmitting circuit E C D and which is inductively linked with a secondary circuit A^1. The latter, it scarcely need be stated, may be wound with any desired number of turns, such as will be best suited for the operation of the device designated by M. The receiving transformer is closely attuned to the oscillations of the transmitting circuit, so that, irrespective of the length of the conductor E^1 C^1 A^1, the points of maximum potential coincide with the elevated terminal D^1, under which conditions the greatest amount of wave energy may be collected and rendered available in the secondary circuit A^1 for useful purposes.

"To complete this description, it may be stated that when it is desired to operate, independently, a great many receiving devices, by such stationary waves of different lengths, the principles which I have set forth in my British Patent, 14,579 (1901), and in my United States Patents, Nos. 723,188 and 725,605 (1903), may be resorted to for rendering the signals or quantities of energy intended for any particular receiver or receivers non-interfering and non-interferable.

"In the above, I have briefly outlined my discovery and indicated only a few uses of the same, but it will be readily seen, that it is of transcending importance for the advancement of many arts and industries, new and old, and capable of innumerable valuable applications.

"Having now particularly described and ascertained the nature of this invention and in what manner the same is to be performed, as communicated to me by my foreign correspondent, I declare that what I claim is:—

"1. The improvement in the art of transmitting electrical energy to a distance which consists in establishing stationary electrical waves in the earth, as set forth.

"2. A system in accordance with Claim 1 which consists in producing in the natural conducting media, stationary electrical waves of predetermined length and operating thereby one or more receiving devices remote from the source of energy and properly located with respect to the position of such waves as herein set forth.

"3. The improvement in the art of transmitting electrical energy, which consists in producing in the earth stationary electrical waves of different lengths, varying their lengths, and causing thereby a resultant wave or effect to travel with the desired velocity over the earth, as above described.

"4. The method of producing effects at a distance, which consists in impressing upon the terrestrial globe stationary electrical waves, varying their characteristics and relations, and causing thereby corresponding effects in distant receivers, as above described.

"5. The improvement in the art of transmitting and distributing electrical energy, which consists in producing in the terrestrial globe intersecting trains of stationary electrical waves, establishing thereby regions of definite electrical activities, and collecting the energy, as above set forth.

"6. The method of producing electrical effects increasing with the distance, which consists in impressing upon the earth electrical oscillations of a frequency of about six per second, and of such character as to give rise to a stationary electrical wave, as set forth.

"7. The method of producing great electrical movements in the terrestrial globe, which consists in rendering it resonant by impressing upon it electrical waves of definite length and duration, as above specified.

"8. In the system as hereinbefore described for the transmission of electrical energy, generating apparatus adapted for producing a resonant condition in the terrestrial globe, as above specified.

"9. In the system as hereinbefore described for the transmission of electrical energy, a transformer adapted for the production of great electrical movements in the terrestrial globe, as above specified.

"10. In the system as hereinbefore described for the transmission of electrical energy, a source of primary electrical oscillations such as a condenser circuit and a secondary circuit inductively linked with the same and adapted for throwing the terrestrial globe into resonance, as above specified."

If Mr Tesla's statement that the fundamental electrical vibration of the earth has a frequency of six per second is founded directly on experimental evidence, and not merely on theoretical considerations, it is a proof that the earth is surrounded by a conducting shell, for if this were not so its lowest frequency of vibration would be seventeen per second, as mentioned by Professor Fitzgerald.

We can also make an estimate of the height at which the conductive layer commences, for calculation shows that since a height of 60 miles corresponds to a frequency of 10 per second, and a height of 6 miles to 3 per second, a frequency of 6 requires a height of about 35 miles, which is the same as that found from physical data in Chapter XVII.

Thus we should have an entirely independent proof of the existence of a conducting layer in the atmosphere, and of the fact that the dielectric, with which we have to deal in wireless telegraphy, is a layer of air about 35 miles thick between two concentric spherical conductors.

Professor Fessenden has noticed in working between America and Scotland that there are apparently two impulses received for every spark, and has suggested that the second may come the longer way round the earth. The interval between them is approximately a fifth of a second, which would indicate the height of the upper conducting layer to be about 30 miles.

CHAPTER XIX.

ADJUSTMENTS, ELECTRICAL MEASUREMENTS, AND FAULT TESTING.

Continuity of Conductors.—If the conductor is a closed or nearly closed circuit, its continuity may easily be proved by means of a cell and galvanometer of any type, or an electric bell; but if it be an open circuit such as an aerial wire, in which the conductor terminates in an insulating medium, other methods are more convenient. Aerial and earth wires are frequently covered with insulating material, a break in the conductor may therefore exist which is not visible on account of the insulation. If the broken ends of the conductor be only a small distance, say 2 or 3 mm. apart, there will be no difficulty in transmitting signals, but it will be impossible to receive any. If, therefore, you find that though your transmitting gear and receiver appear all right, you can get no answer from the other station, do not conclude that it is necessarily "his" fault, it may be your aerial.

A simple test, in such a case, is to hold the buzzer (an electric bell with the gong off, being a producer of electric jigs) near the aerial, and move it slowly along the aerial away from the receiver to which the aerial is kept connected. The receiver will respond to the buzzer as far as the aerial is continuous. It is most likely that, if a break does occur, it will be at a place where the wire is subject to repeated bending and unbending by the wind or other cause.

Insulation.—The first thing to note in wireless telegraphy is, the insulation does not insulate *per se*. That is to say, that a piece of insulating material, whose resistance to continuous currents may be millions of megohms, may be perfectly capable of transmitting the whole aerial wire current of a wireless station. It all depends on the shape.

If the insulator is a comparatively thin sheet between conducting plates the high frequency currents will pass through it, as a dielectric current, with as little loss as if it were a thick conductor. Thus what is usually known as a "condenser" may form part of any circuit which is intended to conduct high frequency currents only. Even a condenser of very small capacity such as may be made by twisting two short pieces of insulated wire together is not negligible and may transmit an appreciable current. I may mention two important cases. The first is, geometrically, on a large scale, and the second on a small one, their electrical dimensions are, however, not so very different. If a large sheet of conducting material, say wire gauze, or a metal cylinder a few feet in length and a foot or so in diameter, be suspended on insulators a foot or two above the ground, the condenser formed is capable of transmitting large currents, such as are used in transmission, if the frequency be high.

The second case illustrates what may occur in the receiving circuit of a station. If two insulated wires lie, or are wound side by side throughout a few inches, or feet, of their length, sufficient current may pass between them to actuate the receiver. Not, of course, a leakage conduction current but a dielectric displacement. In many problems in wireless telegraphy this has to be taken into account, but more particularly in the design of receiving transformers. It is easy to show that for instance an ordinary anti-inductive resistance need not act as a continuous conductor at all, and may conduct the current as well if the bight of the wire be cut as when the wire is continuous.

Insulation in wireless telegraphy means, then, not only high ohmic resistance but extremely small capacity between the insulated wire and the surrounding conductors, whatever they may be.

In ordinary language insulators must be very *thick*, *i.e.*, the distance between the conductors which are separated by the insulator must be large as compared with their areas.

Surface Leakage.—It is also important that any path from one conductor to the other over the surface of the insulator should be long, if high voltages are used, for otherwise sparks may travel along the surface and constitute a leak, even though the insulator is not pierced. In this connection it should be noticed that it is of little benefit to use a hooded insulator if the hood is merely screwed on, an uncovered rod of the same size as the centre piece would be quite as efficient. The spark travels up the film of air on the screwed part of the rod and thus through the hood without difficulty. Much useful information on this subject is contained in Mr Martin's book on Tesla's researches.

From what has been said it will be seen that ordinary insulation tests are useless, the only practical method being to subject the insulator to the working jig-current under the most disadvantageous circumstances as regards damp and dirt that are likely to occur in actual working.

It is often useful to wash outside insulators with fresh water to remove incrustations of salt, sand, and other materials which tend to reduce the insulating power of the surface.

Insulation of Transformers and Induction Coils.—As in the case of outside insulators, the breakdown of an induction coil or high tension transformer is usually of the disruptive kind. A hole is pierced through the insulation and a spark passes intermittently, whenever the potential rises high enough. By earthing alternately the ends of the secondary the position of the fault may often be roughly determined, as the longest spark will be obtained when the faulty end is connected to earth. As a rule the secondary must be taken to pieces and the leak discovered by examination. A temporary repair may sometimes be effected if necessary with paraffin wax (candles) and paper. Leaks of this kind are often due to a flaw in the ebonite or other insulating material, such as an air bubble or a particle of metal. It is therefore best to use oil, from which the air has been extracted, as an insulator, merely trusting to the solid insulator to keep the conductors at their proper distance apart. The breakdown may be largely due to heating owing to dielectric hysteresis, and this is much greater in solids than liquids. In addition, the liquid, even if pierced by a spark, is self-sealing, thus no precaution need be taken beyond slightly reducing the working voltage.

Receivers.—The testing of a receiver consists in (1) making sure that all its parts and connections are made as intended; and (2) in adjusting it on the aerial by the actual reception of signals from the sending station.

In tuned stations, *i.e.*, where the reception of signals depends on the receiving circuits having a definite natural frequency of vibration, it is usually necessary to adjust the aerial circuit to the proper frequency as a transmitter before inserting the receiver (see below).

So many detectors are now in use that it is impossible to go into the adjustments of them all, and indeed nothing but practical work can give a workmanlike knowledge of such a subject. I may mention, however, that it is frequently the subsidiary parts, such as relays or inkers which give the most trouble, the distinctively wireless instruments, even the coherer, are usually more uniform in their action. This is, no doubt, one reason why detectors which work in conjunction with only a telephone receiver are now in so general use. In these receivers there is only the detector itself to adjust, and in many cases this is an exceedingly simple and practically permanent adjustment.

The testing of filings coherers has been gone into in Chapter IV.; no similar data appear to have been published, if, indeed, experiments have been made, in regard to any other form of detector.

Transmitters.—In most stations nowadays the most important adjustment is the equalisation of the frequencies of the aerial and condenser circuits, so that revibration may be obtained. This is usually done by putting a measuring instrument, say a thermo-ammeter between the aerial and earth, and then varying the inductance or capacity until a maximum current is obtained. The instrument may then be removed, if desired, and an equivalent conductor inserted in its place. The action of one circuit on another of the same natural frequency produces, however, not one simple jig, but two of different frequencies and damping. The adjustment of frequencies, then, involves the finding of the greatest maximum current in the transmitting aerial, and the greatest maximum in the detector circuit. It should be remembered we are in general dealing with only three entirely distinct conducting circuits, the condenser circuit of the transmitter, the continuous or nearly continuous conductor formed by the aerials and the earth and the detector circuit. Each of these circuits is broken as to continuous currents in one or more places (condensers), but each is complete as regards jigs, and is coupled to its neighbour electro-magnetically. The term circuit can, of course, be only applied in wireless telegraphy in its widest sense as implying a system of conductors along which a current impulse can pass directly. An inductive coupling is indirect as regards current, there being a transformation of the current energy in any desired proportion across space at right angles to the current; whereas if a condenser be interposed, the entire current goes straight through it. We therefore do not include conductors coupled inductively as part of the same circuit, but we do consider that a condenser *connects* two conductors.

The first person to use a hot-wire ammeter for the purpose of determining the point of maximum revibration appears to have been Captain Wildman of the U.S.A. Army Signal Corps. A convenient form of instrument is Duddell's thermo-ammeter, which is well suited for finding the maximum in the transmitter, while his thermo-galvanometer, a much more sensitive instrument, will determine the point of resonance of the receiving aerial to the transmitted jigs. There are, of course, many other suitable thermo-ammeters, but instruments as sensitive to jigs as the thermo-galvanometer are certainly not common (see Chapter VI.).

Beyond the tuning there is not much adjustment required in the transmitter. The spark length may have to be varied occasionally, and any moving parts of the apparatus seen to. If an induction coil be used the break will require some attention.

In the Lodge-Muirhead system the actual frequency of one of the circuits is determined by a wavemeter, and the other circuits are adjusted to the same frequency. Similar arrangements are used by other companies.

Interrupters.—The efficiency of an interrupter depends on the rapidity with which it breaks the circuit, on the maximum induction produced in core, and on its regularity in action. The rapidity with which the primary current is quelled determines the rapidity of the change of induction in the core, and hence the voltage attained by the secondary current. It is usual to shunt the spark, in the primary circuit, by a condenser, thus allowing the current to become oscillatory, and therefore to reach a zero value long before it would have otherwise done so.

Lord Rayleigh has shown that the condenser is not absolutely necessary to efficient working, if only a sufficiently rapid extinction of the primary current can be otherwise ensured. His plan, though perfectly satisfactory as a demonstration, was not exactly applicable to wireless telegraphy; in fact, he fired a rifle bullet at the primary wire. A wide gap was cut in the wire, and in an extremely small fraction of a second the current was stopped. The velocity of the bullet was probably about 2,000 feet per second—a velocity not attainable by ordinary mechanical arrangements, and therefore impracticable on induction coils. The experiment, however, proved that the rapid stoppage of the current was the essential factor in the production of a large secondary voltage. The best interrupter is therefore that which produces the largest and most rapid change of induction in the core, and which repeats the operation with the greatest regularity.

The adjustments should thus be aimed to keep these facts in view. The contact must be of long enough duration to allow the core to become magnetically saturated, or nearly so, and the break must be quick and regular. The wireless telegraphist does not require a shower of long thin threadlike sparks, but a regular succession of short fat ones, which must crackle sharply and not fizz or flare.

Capacity.—Capacities in wireless telegraphy are usually very small as compared with what have hitherto been considered important. They also, as a rule, have to be reckoned for high frequencies which give different results from constant voltages, and as the stress in the dielectric is not steady but varying rapidly, the dielectric hysteresis of the insulator is not a negligible factor. That is to say, that the stress in the dielectric is not perfectly reversible; some of the energy is wasted and cannot be got back. The effect is like magnetic hysteresis, or like the imperfect elasticity of a solid, which on being bent or compressed does not return to its original shape entirely of its own accord. In all three cases the energy apparently lost is transformed into heat. The quantity of heat produced in each bending and unbending, or each to and fro of potential in the dielectric, is small, but if the oscillations occur many hundred thousand times in a second there is no time for the heat to radiate, and the dielectric may become so hot that it loses its insulating properties and is pierced by a spark discharge. Even much less rapid oscillations, if continued long enough, will break down a thick plate of glass.

The first occasion, as far as I am aware, on which experiments were made with a current of very high voltage, and also considerable horse-power, was at the Crystal Palace Electrical Exhibition in 1892. In Messrs Siemens' experiments an alternating current from an alternator, driven by a 25 H.P. engine, was transformed up to 50,000 volts. Terminals were placed at the middle points of opposite sides of a thick piece of plate glass over a foot square. When the current was turned on, the plate was lit up by a magnificent display of seaweed-like splashes of sparks running from each terminal over the central parts of the plate. In a very short time, however, the display ceased, and a small white spark was visible passing right through the plate between the terminals. On examination, after the current had been turned off, it was seen that a small hole had actually been melted through the glass. There was no crack, but a slight burring up of the glass round the hole, which itself was like the bore of a fine capillary tube. This showed that it was not mere mechanical strain which had caused the perforation, but that the glass had actually been heated by the continued alternating stresses until it had melted. Glass when heated rapidly loses its dielectric strength, hence after the central portion of the plate had become warm the current would rapidly concentrate itself and increase the heating until a puncture took place.

It is obvious that dielectric hysteresis such as this would cause serious loss of energy in a powerful wireless station, where the transmitting apparatus greatly resembles that used in the experiments quoted, the chief difference being in the frequency of the jigs in the condensers, those in the wireless apparatus being, of course, of much the higher frequency. It has therefore been found advantageous in many cases to employ a dielectric such as air or oil, in which the hysteresis is practically inappreciable. Air under a pressure of 70 or 80 lbs. per sq. inch will withstand much higher potentials than at atmospheric pressure, and has for some years been used to increase the insulation, and therefore the available output of electrostatic machines. More recently it has been utilised by Fessenden and others to decrease the sparking distance for a given voltage in wireless stations, and Fleming and others have proposed its adoption as a suitable dielectric for condensers. It seems possible that where initial expense is not of the first importance, compressed air may take the place of other dielectrics in condensers through which a large dielectric current of high frequency has to be passed. Otherwise oil, particularly resin oil, carefully freed from air bubbles, is probably the best available insulator.

Mr Tesla, whose enormous experience in the use of high frequency currents gives great weight to his advice, says that in using a condenser in connection with a high frequency coil, "It should be an oil condenser by all means, as in using an air condenser considerable energy might be wasted." (It should be noted, however, that he means air at atmospheric pressure.) He remarks also that he generally uses linseed or paraffin oil, as there is apparently little difference, and insists very strongly on the benefit of excluding all gaseous matter on account of the heating produced by molecular bombardment when high voltage and frequency are used.

In spite of these advantages of oil, glass condensers, of the Leyden jar type, are in very general use in wireless telegraphy ; their lightness and portability being much in their favour. Rather thin glass is used as a rule, as it is found to be more reliable, and when high voltages are required, a couple or more of the jars may be put in series, thus reducing the voltage on each.

For condensers in the receiving circuit no such precautions against high voltages are necessary, and a thin layer of waxed paper or any similar dielectric may be used.

It must be remembered that the currents in the transmitter condensers sometimes attain very large values, occasionally hundreds of amperes; for this reason it is advantageous to use plates whose conductance is somewhat better than that of the old-fashioned tinfoil sheets; thin copperfoil serves the purpose well, and though it is rather more expensive and less easily worked into shape, the terminal connections are much more satisfactory and less liable to go out of order.

Fault-Testing—Condensers.—As in other insulation tests mere megohms are not sufficient. An indefinite number of these might be found lying in a condenser with a punctured dielectric, perfectly useless for wireless purposes, if only a low-testing voltage (say 1,000 volts) were applied. It is essential to use at least the working voltage and frequency, and then to observe results. If there is a spark-gap in the testing circuit, the character of the spark or its absence will indicate whether the condenser is performing its function properly. Probably the best test would be to put the doubtful condenser in parallel with an adjustable spark-gap, and then increase the latter from nearly zero, *i.e.*, spheres nearly in contact, up to the required spark length. The voltages corresponding to a large range of spark lengths will be found in Tables X. and XI. It should be remembered that in revibrating circuits the voltages are occasionally really very high, even for wireless telegraphy. For instance, Professor Fleming speaks of sparks occurring at the top of an aerial, of a length which indicated a potential of seven million volts, when the aerial was excited by a closed jig circuit of equal frequency.

Condensers — Measurement of Electrical Dimensions.—There are many well-known methods of finding the capacity of a condenser when a steady voltage is applied. For instance, Lord Kelvin's method of mixtures. Descriptions and formulæ for such tests may be found in pocket-books of electrical rules and tables, such as Munro and

Jamieson's, and in most text-books. Low frequency measurements do not, however, give accurately the high frequency capacity, hence it is necessary to employ high frequency currents when measuring capacities used in wireless telegraphy.

Professor Rutherford, in 1896, made a number of determinations of the dielectric constant of various materials at high frequencies, using his magnetic detector as a current measurer. The results showed very considerable variations from ordinary steady voltage measurements. His method is of course applicable to the determination of the capacity of condensers.

Other methods, which are perhaps more suitable for the measurement of very small capacities, are based on the principle of Lodge's resonating Leyden jar circuits (see page 10). If the two circuits be tuned to one another, it is obvious that the equality of two capacities may be proved by putting first one and then the other in one of the circuits. If nothing else has been changed, the capacities must be equal if they give equally good resonance sparks. Thus any number of equal condensers might be obtained by the various combinations of which, in parallel and series, a standard of any desired value might be built up and used for the measurement of any unknown capacity. It is also clear that if the inductance and capacity of the primary circuit and the inductance of the secondary were known, and could be given definite values at pleasure, it would be possible to measure directly any capacity within the range of the instrument. Several instruments on this principle have been constructed, a convenient form being that designed by Professor Fleming, which, though it is mainly intended for the determination of jig frequencies, is suitable for the measurement of capacities such as are common in our branch of telegraphy. The instrument has been named a cymometer. We shall describe its use in considering the measurement of wave-lengths. The law which connects the capacity and inductance of a condenser circuit with its natural frequency of vibration is a simple one, at least to a first approximation, which is good enough for most purposes. If n be the frequency, or number of complete periods per second, while C is the capacity and L the inductance, in C.G.S. units, the equation $n = \frac{1}{(2\pi\sqrt{CL})}$ holds good. Thus if any two of these quantities be known, the third is easily calculable. We shall return to this later.

Remarks on the Adjustment of Transmitters by Messrs Duddell and Taylor.—By kind permission of the authors I quote the following from their paper read in 1905 to the Institution of Electrical Engineers :—

" To carry out successfully measurements of the strength of oscillations set up in a receiver antenna under the various conditions set out in this paper, necessitates very great care in the working of the transmitter to avoid irregularities of action. In spite of all precautions some variations were unavoidable, but they were not of such a nature as would invalidate results obtained if other conditions remained constant. To produce the most regular and uniform strengths of oscillations in the transmitter wire it was necessary to give attention to the following points :—

" 1. To use an interrupter which is capable of uniform and consistent operation.

" 2. To use a definite adjustment of the spark or discharge gap in relation to the power applied. This gap must not be too short, or erratic multiple sparking is obtained; nor must it be too long, or frequent " misfires," or spark failures then result.

" 3. To use sparking knobs of a size suited to the capacity and length of spark adopted.

" 4. To keep the insulation of all parts of the high tension system as nearly perfect as possible, and minimise all brush discharges.

" 5. To use Leyden jars with good metallic coatings The nitrous fumes developed during use cause rapid deterioration of unprotected tinfoil coatings. The jars used were completely varnished over the inside coatings with shellac varnish, except where connections were made.

"6. To keep the height and relative positions of the antenna wires, earth leads, and earth netting, constant.

"The most difficult points to negotiate were those connected with the adjustment of the spark-gap, as, having taken all precautions, sparks would occasionally misfire, or vary in intensity. Owing, however, to the method adopted in observing and recording the reflections obtained on the receiving instrument, occasional 'misfires' did not necessarily affect the readings, because for each reading a continuous series of sparks, lasting half a minute, was used, and any erratic momentary deviations in the deflections were not considered in recording the mean deflection, the steady value only being taken into account. Apart from this effect mysterious changes in the spark were often observed. When the transmitter oscillations, as indicated on the thermal ammeter, were below par, it would often be possible to resuscitate the sparks by slightly moistening the discharge knobs. It was also found that for relatively large capacities, such as that used for wave No. 3, it was impossible to get uniform results with 1-inch diameter brass knobs on the discharger, when using a short gap. If, however, the 1-inch knobs were replaced by $\frac{1}{4}$-inch brass rods with rounded ends, a much more uniform effect was obtained for short sparks of, say, $\frac{1}{8}$ to $\frac{1}{4}$ inch, but for longer sparks the knobs were preferable. On the other hand, with the smaller capacities used, the knobs were found suitable under all conditions of spark lengths adopted in the experiments, whilst the rods gave poor results.

"As regards the form of interrupter, the most consistent results were undoubtedly given by the mercury turbine, if run at speeds less than about 1,500 revolutions per minute. At higher speeds there is a tendency to churn up the alcohol and mercury, forming a mercury emulsion, which seriously interferes with the action of the interrupter. It is also very important to keep the resistance of the primary circuit through the spark coil as low as possible, and adjust the current strength by regulating the voltage applied rather than by inserting resistance. The Grisson interrupter gives a much more rapid series of sparks, however, and therefore produces a large deflection on the receiver."

Relays.—A relay of one kind or another is used in many systems of wireless telegraphy; indeed we might include all detectors whether coherers or otherwise under the head of relays, for their function is essentially that of a relay. However, since it is convenient to deal with them separately, we shall limit the term to the older type of electro-magnetic relay. At the same time it should not be forgotten that all detectors are in reality jig relays, *i.e.*, they are actuated by the electric jig and turn on the current of a local battery.

The ordinary telegraphic relay consists of an electro-magnet, the windings of which are connected to the line from the distant station. As the line current increases the magnetisation of the electro-magnet, a ditton, or tongue, of soft iron, is pulled over and makes contact with a stop, thus completing the circuit of the local battery. It is clear that in the adjustment of a relay we have to deal with two most difficult experimental problems. Firstly the ditton must move sharply over under the influence of a very slight change in the magnetic field, and it must return as rapidly when the field returns to its normal value. It is therefore not possible to gain sensibility by making the controlling forces such that the ditton is in nearly unstable equilibrium; for if that were done the return movement would be uncertain and slow. The best working conditions with most relays appear to be attained when the tongue is held back by a somewhat stiff helical spring, and has therefore only a very small range of motion altogether.

If a relay is to act rapidly it is clear that the mass of the moving part must be small, and that the controlling spring must be a comparatively stiff one. The reason for a stiff spring is that otherwise it is very difficult to set the contacts very close together, and still avoid disturbance from accidental vibration. With a stiff spring the gap may be almost microscopic, and yet, since the motion only stretches the spring a very little, the sensibility may be quite as great as with a weak one and a larger gap, for the controlling force exercised by the spring is just proportional to the amount of its extension. Under these circumstances the contact is quickly made, and at the same time the action of the relay is steady and reliable.

The other problem in connection with relay adjustment is the old one of how to maintain a *clean surface*. The contacts cannot but be surfaces of some sort, usually they are *interfaces* between metal and air. Now a surface is thin, very thin (see Euclid's well-known work on geometry), so it does not take much material to alter it. Hence the difficulty of obtaining anything like constancy in the action of the contacts of a relay. A microscopic speck of dust may prevent the circuit being closed, and to remove one speck of dust without putting two others in its place is not so easily done as might be expected.

A good relay therefore must be dust-proof, and should never be opened in a dusty room. Its contacts may be cleaned, if necessary, by placing a small piece of fine cardboard or smooth paper between the contacts and pressing them together, at the same time turning the paper to and fro. The use of condensers and shunts in parallel with the gap between the contacts has greatly increased the length of time that a contact will keep clean, and thus has improved the reliability.

Besides the more ordinary types of relay, the best of which will work with about one ten-thousandth of a watt, there are a number of special relays of great sensitiveness which have been designed for use on long distance submarine cables. They are usually of the moving coil type, but in general are troublesome to adjust and not by any means portable. This year (1906), however, Mr H. W. Sullivan has invented a relay which while having extraordinary sensibility, working, in fact, with about one ten-millionth of a watt (one volt through ten megohms), is portable and easy to adjust. He has designed the instrument specially for wireless telegraph work, more particularly to act as a call relay in stations where noncontinuous detectors, *e.g.*, detectors of the electrolytic type, are used. It has been tried and proved most successful, at more than one long distance station. Mr Sullivan has kindly permitted me to reproduce the following description * of the construction and action of this relay:—

"With the extension of the distances over which signals are sent at sea, necessitating the use of very sensitive apparatus with which signals can only be satisfactorily received by means of the telephone, a great want has been felt for an apparatus which will give a clearly audible call signal when working at great distances. In order to meet these conditions, Mr H. W. Sullivan has designed a specially sensitive call relay for use with electrolytic coherers and similar wireless detectors.

"The instrument consists essentially of a very sensitive moving-coil galvanometer, of a form somewhat similar to Mr Sullivan's well-known marine galvanometer, to which contacts are fitted. A specially strong magnetic field is obtained by permanent magnets, and some idea of the sensibility may be obtained from the fact that with one dry cell working through eight megohms the limit of working has not been reached. This is a great improvement on most of the relays at present on the market, which as a rule will not work satisfactorily with a less current than that given by one cell through 250,000 ohms. The success of such an instrument as a relay naturally depends on the material, design, and method of adjustment of the contacts. The material is a special alloy, the outcome of many years of experience in this class of work. If both the fixed and moving contacts were mounted on rigid arms trouble would be experienced in the moving arm rebounding and never making more than a momentary contact insufficient to ring an ordinary trembler bell. Mr Sullivan gets over that by not only mounting the 'fixed' contacts on springs, but by also making the light moving arm of springy material. When the position and strength of the springs is correctly adjusted, both yield slightly when the arm strikes the contact, and this not only diminishes the rebound by softening

* From the *Electrician*, by kind consent of the proprietors.

the blow, but also prolongs the contact by following up the return movement. A very neat adjustment is provided for the fixed springs, one screw shifting a slider which alters the effective length of the spring, and another swinging the carriage on which the spring is mounted. A contact system is provided at each side, so that it is immaterial which way current is sent through the instrument. Another interesting detail is the provision of a stop limiting the return movement of these springs, so that all possibility of sticking is avoided.

"The moving coil is mounted between top and bottom suspensions of phosphor bronze strip, and is fully damped electro-magnetically, and balanced for use on board ship. As in the case of the Sullivan 'speaking' galvanometer for submarine cable work, the frame carrying the coil suspension and relay contacts can be bodily removed by sliding out vertically, and as the connections to both the line and local circuits are made by stout springs bearing on pillars, a new coil can be substituted in a few seconds in case of a suspension becoming defective, or a coil otherwise damaged. The whole instrument is simple, strong, and not liable to be affected by vibration, notwithstanding its high degree of sensibility.

"Although primarily designed as a call relay, the apparatus would seem capable of recording signals at slow speeds at distances considerably in excess of those already accomplished with recorded signals, but experiments in this direction have not yet been carried out. In addition to the relay contacts the instrument is provided with a mirror, so that if desired a visual signal can be made by the movement of a large rectangular 'spot' of light on a wall or bulkhead of the ship. The movement of a large spot of this kind would be visible to all in the room without close watching, and a useful check would be provided in the case of an accidental interruption of the local bell circuit. Mr Sullivan is a firm believer in the advantages of the 'electrolytic coherer' over other detectors, and is perfecting several pieces of apparatus designed to work with this system."

An interesting type of relay, not as yet used in wireless work, is the magnifying relay invented by Taylor and Dearlove. It consists of two movable coils mounted on one axis, the lower one being in the field of, and being actuated by the line current, and the upper one in a powerful alternating field locally produced. The upper coil is connected with a rectifier. When no current is flowing in the lower coil the upper stands with its plane parallel to the lines of magnetic force. When a current turns the lower coil, the upper turns with it and therefore allows lines of force to pass through it, and as these are alternating, a considerable alternating current is generated by induction in it. This current is passed through the rectifier and from thence through the recording instrument. The instrument is most ingenious, and does away with contacts, but is unfortunately rather complicated.

Measurement of Frequency and Wave-Length.—Since the frequency of oscillation of any circuit or conductor depends upon its inductance and capacity, it is possible by determining these to obtain its value. Various instruments are now made in which a variable and graduated condenser is combined with an adjustable inductance, so that the oscillation frequency of the circuit of the instrument can be obtained from the readings. The instrument is then placed so that a part of its circuit is in proximity to the circuit whose frequeney it is desired to measure. A jig current is then started in the undetermined circuit, and the instrument is adjusted until a thermo-galvanometer or other indicator on it shows a maximum induced current. This occurs when the natural frequencies of the circuits are the same, it is therefore only necessary to calculate this frequency from the values of capacity and inductance shown by the instrument.

In one type of instrument designed by Professor Fleming, the movement of one handle varies the capacity and inductance proportionately ; it is therefore possible to graduate the instrument, which Professor Fleming has named a cymometer, directly in terms of frequency. This is a convenience for rapid determinations, but limits, in some respects, the general utility of the instrument; it is well suited, however, for the measurements usually required in wireless telegraphy. Fig. 129 shows the instrument diagrammatically, and except in regard to the mechanical connection of the capacity with the inductance, and the neon tube used as an indicator, the diagram would apparently be also a correct representation of the wavemeters used by other companies.

An instrument of this kind may be used, of course, to determine any one of the quantities in the equation

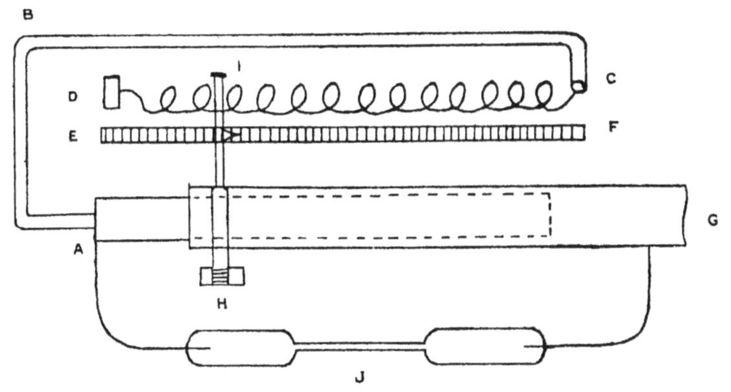

FIG. 129.—FLEMING'S CYMOMETER.

A B C, Thick Copper Conductor; *CD*, Helix used as Variable Inductance; *EF*, Scale; *AG*, Co-axial Tubes forming Variable Condenser; *HI*, Slider which varies Inductance and Capacity Simultaneously; *J*, Vacuum Tube to indicate Maximum Resonance.

$n = \frac{1}{2\pi\sqrt{CL}}$, provided the other two are known. For instance, if the capacity of a condenser has to be found, it is only necessary to put it in circuit with a standard inductance, and then determine the frequency of the circuit so formed by means of any wavemeter. An unknown inductance may likewise be measured by putting it in series with a known capacity. This is in fact probably the most accurate way of determining either quantity, as measurements of capacity taken at steady voltages are not correct for jigs, and the inductances used in wireless telegraphy are as a rule so small that measurements by other methods are very difficult.

If a helix be long in proportion to its diameter, its inductance may be approximately determined by the formula $L = l(\pi dn)^2$; where L is the inductance, l the length of the helix, d its diameter, and n the number of turns per unit length. For instance a helix of 5 cm. diameter, 10 cm. long, and 2 turns per cm., would have an inductance $L = 10(3.14 . 5.2)^2 = 10 . 1{,}000 = 10{,}000$ C.G.S., or 1,000 per cm. of its length, approximately.

The capacity of a condenser formed of two concentric metal cylinders, which is a convenient form in which to make a variable standard, is given by the equation

$$K = S\frac{l}{2\log_e \frac{R}{r}}$$

where K = capacity in C.G.S. electrostatic units (these may be converted to microfarads by dividing by 900,000); R and r are the radii of the external and internal cylinders respectively in cms.; S is the specific inductive capacity (dielectric constant) of the insulator between the cylinders; l is the length in cms. of the part of the cylinders which overlap—this, in the case of a variable condenser in which the variation is produced by drawing one cylinder partially out of the other, is subject to a small correction.

Thus the capacity of two cylinders overlapping one another throughout 50 cms. of their length, whose radii are 2 and 2.2 cms., separated by an ebonite tube (dielectric constant = 3), with walls .2 cm. thick, is approximately—

$$K = \frac{50 . 3}{2\log_e\left(\frac{2.2}{2}\right)} = \frac{150}{2 \times .0953}.$$

$K = 778$ C.G.S. electrostatic units.

$= \frac{778}{(3 . 10^{10})^2}$ C.G.S. electro-magnetic units.

$= 0.00087$ microfarad.

The frequency of a circuit containing no appreciable capacity or inductance beyond those of the above condenser and helix would be—

$$n = \frac{1}{2\pi\sqrt{.00087 \times 10^{-15} \times 10^{4}}}$$

$$= \frac{1}{6.28\sqrt{87 \times 10^{-16}}}$$

$$= \frac{10^{8}}{6.28 \times 9.34}$$

$$= 1.71 \times 10^{6};$$

or nearly two millions per second.

In the Fleming cymometer the frequency and wave-length may be read directly from the scale. In wave-meters in which the condenser and inductance are not simultaneously varied, a short calculation like the above is necessary.

It is clear, then, that the frequency of any circuit may be calculated if its capacity and inductance are known, or it may be found by resonance with a circuit of known capacity and inductance.

Damping.—The oscillatory current produced by the sudden breakdown of the resistance of the spark-gap when the spark occurs gradually dies out, partly through the conversion of the energy stored in the charge into heat, and partly by radiation. By radiation we do not mean, in this connection, only straight line radiation through free space, but also the oscillatory currents which spread out over the earth's surface. These also are finally damped out by the resistance of the earth and the surrounding air.

The rate of damping of the jig in a transmitter is, of course, a very important factor, and its determination is necessary before the efficiency of the installation can be known. In this country the damping is usually defined as the ratio of the amplitude of the second half-oscillation to the amplitude of the first half-oscillation, *e.g.*, of the first negative to the first positive wave if we begin with a positive. Abroad the number usually expresses the ratio of the second positive to the first positive, *i.e.*, the loss by damping is reckoned over the complete oscillation. The Napierian (or hyperbolic) logarithm of the reciprocal of the damping, *i.e.*, of the ratio First Amplitude: Second Amplitude, is called the logarithmic decrement of the motion.

Several experimental methods are available, of which we shall describe one. Professor Rutherford's method of determining the coefficient of damping of an electrical oscillator is as follows:—

The oscillatory currents are made to pass round a very small helix of fine wire in the axis of which is placed a small steel needle or bundle of fine steel wires, which has been magnetised to saturation. The deflection of a suspended magnet or magnetometer caused by the steel needle when placed at a given distance is noted. The needle is then placed in the helix, and the oscillatory current passed through the latter. The change of magnetisation of the needle is noted by an observation with the magnetometer as before; call this result (A).

The needle is then remagnetised to saturation, and after a measurement with the magnetometer, is placed in the helix in the reverse direction. A spark is passed, and another measurement (B) of the change of magnetisation made. These changes in magnetisation are proportional to the amplitudes of the first and second half waves, hence from these observations the damping and logarithmic decrement may be calculated. There are, of course, other methods, but this is one of the simplest to carry out, and appears to give very satisfactory results.

Lieutenant Tissot,* using a wavemeter with constant

---

* *Soc. Int. Elect. Bull.* 6, July 1906.

inductance and variable capacity, first determines the frequency N of the given circuit. Leaving the wavemeter equifrequent with the circuit under test he next introduces non-inductive resistances (short straight wires from 0 to 3 ohms) in its circuit and observes the reading of the thermogalvanometer corresponding to each resistance. On plotting the resistances as abscissæ and the currents as ordinates a straight line is obtained, which on being produced cuts the axis of X in some point whose abscissa is, say, x on the scale chosen. By multiplying x by $1/2LN$, where L is the inductance of the wavemeter, the value of $\delta/2+\delta'$ may be obtained; δ being the logarithmic decrement of the main circuit and δ' that of the wavemeter. The latter is usually small in comparison with the former and may generally be neglected; hence $x/2LN$ is approximately equal to δ, the decrement required.

CHAPTER XX.

ON THE CALCULATION OF A SYNTONIC WIRELESS TELEGRAPH STATION.

THIS chapter is almost entirely a translation Signor Alfredo Montel has very kindly permitted me to make of a paper contributed by him to *L'Elettricista* of 15th June 1906, and gives a concise description of the calculations involved in the design of a wireless telegraph station. The methods and results are, of course, only roughly approximate, as exact data and theory are still wanting. The approximation, however, gives an excellent idea of the magnitude of the quantities involved, and of the proper dimensions of the apparatus in any given case. I have added a few short explanations where they seemed necessary, but have otherwise adhered closely to the original.

It should be noticed that, contrary to usual English custom, the frequency n represents the number of half periods per second. To reduce the equations in which it occurs to English notation, we should therefore have to write $2N$ in place of n, N being the number of *complete* periods per second. In the numerical examples it is simpler to make the calculations as they stand, recollecting that we should call the frequency one and a half and not three millions per second.

Without further prologue we commence Signor Montel's paper :—

"The transmitting station consists essentially (Fig. 130) of a closed oscillating circuit, and of an antenna coupled to the same. C_1 is the condenser, BB a source of energy (*e.g.*, the secondary of a transformer), FF the spark-gap, L_1 the inductance of the circuit.

"In syntonic telegraphy large quantities of energy occur, hence if C_1 is to be of convenient size, while C_1L_1 does not exceed a certain limit (and also to economise energy), it is necessary to make L_1 small. The ohmic resistance should be small, and the gap FF should be subdivided into small

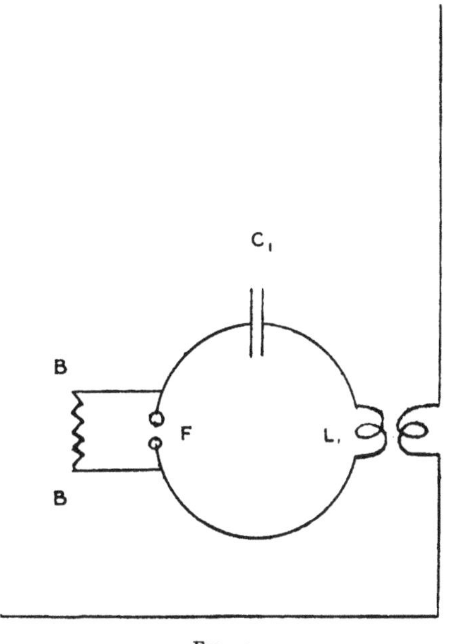

FIG. 130.

parts. These conditions are chosen with a view to making the coefficient of damping as small as possible.

"The decrement (damping referred to the number of periods) depends mainly on the length of the spark, which may be from .4 to .5 cm. according to the capacity of the condenser, &c., and the resistance of the circuit. It depends also on the material and size of the spheres FF, and on the dielectric hysteresis of the condensers. Let α_1 be the decrement in the given oscillating circuit, and let us assume

$a_1 = 0.06$, since this represents as small a decrement as we can expect in a circuit containing a spark-gap.

"In making the calculations for a circuit, one usually starts with the frequency n (the number of reversals per second), or what is equivalent, with the wave-length, which one intends to adopt for transmission. The wave-length should be greater in proportion to the distance, since waves of great length pass more easily round obstacles, and because greater amounts of energy necessitate greater wave-lengths. It does not do, however, to exaggerate the wave-length too much, since by so doing the sensibility to atmospheric disturbances is also increased.

"We have*—

$$n = \frac{1}{\pi\sqrt{C_1 L_1}} \qquad (1)$$

L_1 is given by the form and dimensions of the circuit and the diameter of the wire. If, for example, the circuit is circular and of radius R, and the wire of radius r, we have the equation—

$$L_1 = 4\pi R\left[\log\frac{8R}{r} - 2\right], \text{ C.G.S.} \qquad (2)$$

In this equation the internal inductance of the conductor is neglected since it is hardly appreciable with high frequencies and non-magnetic conductors.

"From (1) we obtain, by algebra—

$$C_1 = \frac{1}{\pi^2 n^2 L_1} \qquad (3)$$

Let V_1 be the potential to which we require to charge the condenser on account of the distance to which we wish to telegraph, since the latter depends on the current in the aerial, and this again on its capacity and potential—

$$w = \frac{1}{2} C_1 V_1^2 \qquad (4)$$

represents the quantity of energy consumed in one dis-

* See remark on frequency at commencement of chapter; also note that C and L are in C.G.S. electro-magnetic units (microfarad $= 10^{-15}$ and henry $= 10^9$ of these units).

charge, and if we take m as the number of these per second,

$$W = \frac{1}{2} m C_1 V_1^2 \qquad (5)$$

represents the activity or horse-power of the station.

"The discharge must be oscillating, with as little damping as possible, *i.e.*, the resistance and radiation must be negligible. If, therefore, i_1 be the current of the discharge, then I_1, its maximum amplitude, is approximately for the case considered in which the damping is small,

$$I_1 = \frac{V_1}{\pi n L_1} \qquad (6)$$

"*Numerical Example — Condenser Circuit of Transmitter.*—If the circuit be circular, the diameter of the wire 0.8 cm., the radius of the circuit 30 cm., and value of $n = 3 . 10^6$, we get from (2) and (3)—

$$L_1 = 4 . 3.14 . 30 \left[\log_e \left(\frac{240}{0.4} \right) - 2 \right] = 1656 \text{ C.G.S.}$$

$$C_1 = \frac{1}{3.14^2 . 3^2 . 10^{12} . 1656} = 6.8 . 10^{-18} \text{ C.G.S.} = 6.8 . 10^{-3} \text{ micro-farad.}$$

If we make $V_1 = 15{,}000$ volts $= 15 . 10^{11}$ C.G.S., and $m = 30$, we get from (4) and (5)—

$$w = 0.5 . 6.8 . 10^{-18} . 15^2 . 10^{22} = 0.765 . 10^7 \text{ C.G.S.} = 0.765 \text{ watt.}$$

$$W = 30 . 0.765 = 22.95 \text{ watt.}$$

$$I_1 = \frac{15 . 10^{11}}{3.14 . 3 . 10^6 . 1656} = 96 \text{ C.G.S.} = 960 \text{ amperes.}$$

Now the aerial must have the same frequency n, considerable capacity, and great power of radiation.

"In place of using a direct earth connection, on account of its liability to variation (except in the case of an installation on board ship) it is better to employ an electrical counterpoise, *i.e.*, a conductor of considerable area, near the earth.

"We shall assume for convenience in calculation that the aerial is a single wire, in which case *

$$n = \frac{3 \cdot 10^{10}}{2l} \qquad (7)$$

l being the length of the antenna (aerial), on the supposition that the counterpoise is equal to the aerial given. From this equation we can get $l = 5 \cdot 10^3$, if $n = 3 \cdot 10^6$.

"For telegraphy to a given distance the magnitude of the current in the aerial practically fixes all the other conditions. The maximum potential on the aerial must not go beyond a certain limit, otherwise the charge would be dissipated by leakage; it also depends on the condenser circuit, and on the degree of coupling between the condenser circuit and the aerial. The relation between current and potential may be given as—

$$I_2 = 2nC_2V_2 \qquad (8)$$

which we obtain from the known formula—

$$I = \pi nCV,$$

by taking account of the sinusoidal distribution of current and potential along the aerial. If we know V_2 and give a definite value to I_2, we obtain C_2. But also,

$$C_2 = \frac{l}{2 \log \frac{2l}{r}} \frac{1}{9} 10^{-20} \text{ C.G.S.} \qquad (9)$$

in which r is the radius of the aerial which we are seeking.

"In the case in which r is not negligible, as compared with the length of the aerial, the frequency n in place of the value given in (2) may be taken, according to M. Abraham, as

$$n = \frac{3 \cdot 10^{10}}{2l\left[1 + 5.6\left(\frac{1}{4 \log \left(\frac{2l}{r}\right)}\right)^2\right]} \qquad (10)$$

* According to Mr H. M. Macdonald the complete wave-length should be five times the length of the aerial, and not four times as here assumed. Experiments show that the intermediate number 4.8 is approximately the true value. (See Fleming, "Electric Wave Telegraphy," p. 559.)

But usually it is unnecessary to take account of the thickness of the aerial wire.

"Taking account only of the fundamental oscillation, we have, according to M. Abraham, the decrement α_2 of the aerial given by

$$\alpha_2 = \frac{2.44}{\log \frac{2l}{r}} \quad \text{- - -} \quad (11)$$

in which we neglect the damping due to heating on account of ohmic resistance. This is usually a correct assumption, as Sommerfeld has shown that the joulean heat waste produces a damping represented by the formula

$$\alpha_j = \frac{1.1}{r} \sqrt{\frac{2\frac{\mu}{\mu_0} 2l}{\sigma}} \cdot 10^{-7} \quad (12)$$

in which σ is the conductance in C.G.S., μ the permeability of the wire, μ_0 that of the dielectric medium, and r the radius of the aerial wire, and this quantity is small.

"The inductance of the aerial is, for high frequencies—

$$L_2 = 2 \cdot 2l \cdot \log \frac{2l}{r} \text{ C.G.S.} \quad (13)$$

"The coupling of the circuit of the condenser to the aerial should be as weak as possible, in order that the secondary may not react appreciably on the primary, and in order to have a maxim efficiency we must make sure that the current "loop" is situated at the base of the aerial.

"Let M_{21} be the coefficient of mutual induction in the case of the given instantaneous value of the current at any point in the circuit. In our case we may put instead, the expression $\frac{4}{\pi} M_{21}$.

"The i_1 oscillating in the condenser circuit produces in the aerial an E.M.F. e_2, such that,

$$e_2 = -\frac{4}{\pi} M_{21} \frac{di_1}{dt} \quad (14)$$

and if E_2 be its amplitude—

$$E_2 = \pi n \left(\frac{4}{\pi} M_{21}\right) I_1 = 4n M_{21} I_1 \quad \text{- -} \quad (15)$$

"*Numerical Example—Transmitting Aerial.*—From (7) we get—

$$2l = \frac{3 \cdot 10^{10}}{3 \cdot 10^{6}} = 10^4$$

To fix the diameter of the aerial let us suppose that the capacity is, for example, 300 cm. (electrostatic units), we then find from (9)—

$$300 = \frac{5 \cdot 10^3}{2 \log \frac{10^4}{r}}$$

from which we get $r = 3$. Therefore our aerial should be a conductor of 6 cm. in diameter. Naturally in practice an aerial consisting of a number of wires would be used.

"From (11) we get—

$$a_2 = \frac{2.44}{\log \frac{10^4}{3}} = 0.3$$

As we said before, the formula (7) is only valid when the diameter of the wire is negligible in comparison with the length. Let us see by applying (10) whether this is so in this case:

$$n = \frac{3 \cdot 10^{10}}{10^4 \left[1 + 5.6 \left(\frac{1}{4 \log \left(\frac{10^4}{3} \right)} \right)^2 \right]}$$

"The expression $5.6 \left(\frac{1}{4 \log \left(\frac{10^4}{3} \right)} \right)^2 = 0.005$, which shows that the frequency n corresponds to $2l = 10^4 \cdot 1.005$, which is very nearly 10^4. We may thus neglect the difference.

"From (13) we get—

$$L_2 = 2 \cdot 10^4 \log \frac{10^4}{3} = 160000 \text{ C.G.S.}$$

and from (12)—

$$a_j = \frac{1.1}{3} \sqrt{\frac{2 \cdot 1 \cdot 10^4}{59 \cdot 10^{-5}}} \cdot 10^{-7} = 0.00021$$

We thus see the damping due to joulean waste is quite negligible, as compared with α_2, which we have determined above.

"*The Jig in the Aerial.*—The aerial is the seat of two types of jig, one being its own and the other that of the condenser. In the actual case we have the same frequency in both, and from theory we find that the two oscillations have otherwise the given initial amplitude and the decrements α_1 and α_2 respectively. The two jigs are compounded and give rise to a single resultant jig.

"Before proceeding we may make the observation that if the maximum intensity of the current at the base of the aerial is known, it is possible to determine the maximum potential at the summit.

"In an oscillating circuit in which the instantaneous values of the current and potential may be stated for any point of its length, we have as expressions for the energy of the circuit the well-known formulæ—

$$\left.\begin{array}{l} w = \frac{1}{2}CV^2 \\ w = \frac{1}{2}LI^2 \end{array}\right\} \quad (16)$$

C represents the capacity of the circuit, *i.e.*, the number of lines of electric induction which start from one half of the circuit and finish on the other half, when the difference of potential is unity; L is the inductance of the whole circuit, while I and V are the amplitudes of the current and potential.

"The expressions (16) represent the same quantity of energy which at one instant, when the potential is a maximum, is entirely electrical, and, at another instant, when the current is a maximum, is entirely magnetic. From this we deduce—

$$CV^2 = LI^2,$$

and recollecting that $n = \frac{1}{\pi\sqrt{LC}}$ we find again the equation (6)—

$$I = \frac{V}{\pi n L}.$$

In the case of our aerial we may suppose that C represents the capacity which is given by the formula (9) for the aerial alone, that is to say, of the half of the internal system, we ought to take account of the sinusoidal distribution of the potential along it, and for this reason put instead $\frac{C}{\pi}$, and since L represents the inductance of the whole system which is given by (13), we must substitute for it $\frac{2L}{\pi}$. In equations (16) V represents the maximum difference of potential between the extremities of the circuit. Hence if we designate by V′ the maximum difference between the summit and base of the aerial, *i.e.*, between an extreme point and the middle of the oscillating circuit—

$$V' = \frac{1}{2}V. \quad - \quad - \quad (17)$$

We therefore obtain from (16)—

$$\frac{1}{2}\frac{C}{\pi}(2V')^2 = \frac{1}{2}\frac{2L}{\pi}I^2.$$

From which

$$V' = I\sqrt{\frac{L}{2C}}. \quad (18)$$

If, instead of the maximum values of the aerial voltage and current, we take the effective or virtual values ϕ, γ, we should get analogously

$$\phi = \gamma\sqrt{\frac{L}{2C}}. \quad (19)$$

"The formula $n = \frac{1}{\pi\sqrt{(LC)}}$ in our case of a sinusoidal distribution of current and potential along the circuit reduces to

$$n = \frac{1}{\sqrt{(2LC)}}. \quad (20)$$

And from (18), (19), (20), we obtain

$$I = \frac{V'}{nL}, \quad (21)$$

$$\gamma = \frac{\phi}{nL}. \quad (22)$$

"To return to the thread of our argument. The oscillating potential ϕ_2 of the aerial is the algebraic sum of the two v'_2 and v''_2, and if V'_2 be the common amplitude

$$\left.\begin{aligned} v'_2 &= V'_2 \cdot e^{-a_1 t} \cdot \cos\frac{\pi x}{2l}\sin(\pi n t - \psi') \\ v''_2 &= V'_2 \cdot e^{-a_2 t} \cdot \cos\frac{\pi x}{2l}\sin(\pi n t - \psi'') \\ \therefore\ \phi_2 = v'_2 - v''_2 &= V'_2(e^{-a_1 t} - e^{-a_2 t})\cos\frac{\pi x}{2l}\sin \pi n t \end{aligned}\right\} \quad (23)$$

where x represents the distance from the top of the aerial.

"From theory * there results, in the case of our aerial, where V_2 is as in (8)—

$$V'_2 = \tfrac{1}{2}\frac{\pi E_2}{\alpha_1 - \alpha_2} \quad (24)$$

$$\left.\begin{aligned} V_{2\,max} &= V_1\frac{M_{21}}{L_1} \cdot \frac{2}{\alpha_2}\left(\frac{\alpha_1}{\alpha_2}\right)^{\frac{a_1}{a_2 - a_1}} \\ &= \tfrac{1}{2}E_2\frac{\pi}{\alpha_2}\left(\frac{\alpha_1}{\alpha_2}\right)^{\frac{a_1}{a_2 - a_1}} \end{aligned}\right\} \quad (25)$$

$$\phi_2^2 = \tfrac{1}{8} \cdot \frac{E_2^2\pi^2}{n} \cdot \frac{1}{\alpha_1\alpha_2(\alpha_1 + \alpha_2)} \quad (26)$$

E_2 is given by (15). From which three formulæ, and taking account of (21), we get the currents

$$I'_2 = \frac{\pi E_2}{2nL_2(\alpha_1 - \alpha_2)} \quad (27)$$

$$\left.\begin{aligned} I_{2\,max} &= \frac{V_1 M_{21}}{nL_2L_1}\frac{2}{\alpha_2}\left(\frac{\alpha_1}{\alpha_2}\right)^{\frac{a_1}{a_2 - a_1}} \\ &= \frac{1\,E_2\pi}{2nL_2\alpha_2}\left(\frac{\alpha_1}{\alpha_2}\right)^{\frac{a_1}{a_2 - a_1}} \end{aligned}\right\} \quad (28)$$

$$\gamma_2^2 = \frac{1\,E_2^2\pi^2}{n^3L_2^2}\frac{1}{\alpha_1\alpha_2(\alpha_1 + \alpha_2)} \quad (29)$$

"At the beginning of the jig I_2 and V_2 have the value zero, then, increasing gradually, they reach the maxima $I_{2\,max}$ and $V_{2\,max}$, and then commence to decrease.

* See J. Zenneck, "Elektromagnetische Schwingungen," p. 586.

" ϕ_2 and γ_2 are the effective values of the potential and current. The deflections of the measuring instrument are proportional to their squares.

" The equation (29) represents the effective value of the current during a single discharge, and γ_2 is proportional to a given quantity of energy per unit time. If in one second we have *m* discharges the quantity of energy per second would be *m* times as great, and hence we should have to put

$$\gamma_2^2 = \frac{m}{8} \frac{E_2^2 \pi^2}{n^3 L_2^2} \frac{1}{\alpha_1 \alpha_2 (\alpha_1 + \alpha_2)} \qquad (30)$$

and analogously—

$$\phi_2^2 = \frac{m}{8} \frac{E_2^2 \pi^2}{n} \frac{1}{\alpha_1 \alpha_2 (\alpha_1 + \alpha_2)} \qquad (31)$$

" *Numerical Example—Transmitting Aerial.*—Let us assume $M_{21} = 200$.

" From (13) we have, as we have shown, $L_2 = 160000$ C.G.S. and from (15)

$$E_2 = 4 \cdot 3 \cdot 10^6 \cdot 200 \cdot 96 = 2300 \cdot 10^8 \text{ C.G.S.} = 2300 \text{ volts.}$$

Substituting now in (24)......(28) we get

$$V'_2 = \frac{1}{2} \frac{3.14 \cdot 2300 \cdot 10^8}{0.3 - 0.06} = 15000 \cdot 10^8 \text{ C.G.S.} = 15000 \text{ volts.}$$

$$I'_2 = \frac{15 \cdot 10^{11}}{48 \cdot 10^{10}} = 2.88 \text{ C.G.S.} = 28.8 \text{ amperes.}$$

$$V_{2\,max} = 15 \cdot 10^{11} \frac{200}{1656} \cdot \frac{2}{0.3} \left(\frac{0.06}{0.3}\right)^{\frac{0.06}{0.3 - 0.06}} = 8.30 \cdot 10^{11} \text{ C.G.S.}$$
$$= 8300 \text{ volts.}$$

$$I_{2\,max} = \frac{8.30 \cdot 10^{11}}{3 \cdot 10^6 \cdot 16 \cdot 10^4} = 1.71 \text{ C.G.S.} = 17.1 \text{ amperes,}$$

and supposing that $m = 30$, we get from (31) and (22)

$$\phi_2^2 = \frac{30 \cdot 2300^2 \cdot 10^{16} \cdot 3.14^2}{8 \cdot 3 \cdot 10^6} \frac{1}{0.3 \cdot 0.06 (0.3 - 0.06)}$$
$$= 10050 \cdot 10^{16} \text{ C.G.S.}$$

Therefore

$$\phi_2 = 100 \,.\, 10^8 \text{ C.G.S.} = 100 \text{ volts.}$$

$$\gamma_2 = \frac{100 \,.\, 10^8}{3 \,.\, 10^6 \,.\, 16 \,.\, 10^4} = 0.021 \text{ C.G.S.} = 0.21 \text{ amperes.}$$

"**The Receiving Circuit.**—Let us now pass to the consideration of the receiving station, which may be

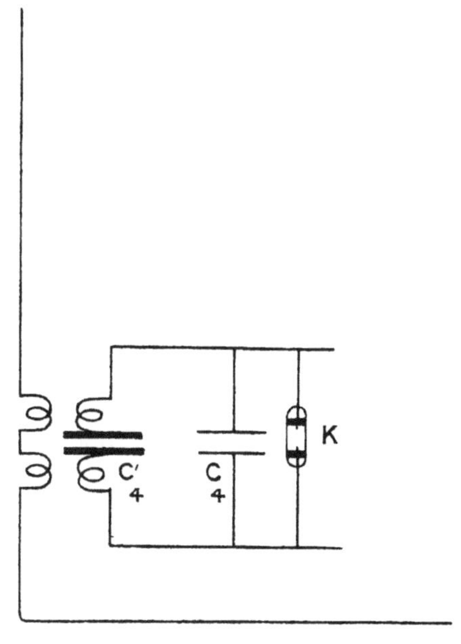

FIG. 131.

arranged as in Fig. 131. The aerial is supposed identical with the transmitting one, and is weakly coupled to an oscillating circuit, containing, for example, a coherer K and two condensers (the Marconi arrangement). The coherer at the first moment of arrival of the jig may be considered as a condenser. C_4 is a condenser of a capacity great compared with that of the coherer, connected in parallel with the coherer so that the general conditions of the circuit may not be affected by the variations of the capacity of the coherer caused by the movements of the filings which it contains. C'_4 is a condenser which serves to prevent a short circuit of the relay and battery. Its capacity is much greater than that of C_4—so great, in fact, that the condition of the oscillating circuit whose ends are at C_4 and the coherer is not altered by its presence. In calculating the circuit approximately we shall not take the capacity of the coherer into account. We have

$$n = \frac{1}{\pi \sqrt{L_4 C_4}}$$

in which n is the frequency adopted by the transmitting station. L_4 varies according to the shape and size of the circuit and the diameter of the wire. C_4 must be, as we said, great enough, but not too great, in order that we may not have to exaggerate the size of C'_4.

"According to the experiments of Duddell and Taylor and others, the effective current in the receiving aerial is proportional to the length of the aerial and to the strength of the transmitting current, and is inversely proportional to the distance between the stations.

"In our case let us suppose that by means of an instrument we know the current γ_3 at the base of the aerial. This current γ_3 is the resultant of a component i''_3 with the decrement α_1 of the transmitting station, and another i'''_3 with the decrement α_2 of the receiving aerial. The equations for potential and current are analogous to (23).

"Knowing γ_3^2 we have from (30) by altering the suffixes

$$E_3^2 = \frac{\gamma_3^2 8 n^3 L_2^2 \alpha_1 \alpha_2 (\alpha_1 + \alpha_2)}{m \pi^2} \quad - \quad (32)$$

Also—

$$V'_3 = \frac{1}{2} \cdot \frac{\pi E_3}{\alpha_1 - \alpha_2} \quad (33)$$

$$V_{3max} = \frac{1}{2} E_3 \frac{\pi}{\alpha_2} \left(\frac{\alpha_1}{\alpha_2}\right)^{\frac{\alpha_1}{(\alpha_2 - \alpha_1)}} \quad (34)$$

$$I'_3 = \frac{\pi E_3}{2 n L_2 (\alpha_1 - \alpha_2)} \quad - \quad - \quad (35)$$

$$I_{3\,max} = \frac{1}{2 n L_2} E_3 \frac{\pi}{\alpha_2} \left(\frac{\alpha_1}{\alpha_2}\right)^{\frac{\alpha_1}{(\alpha_2 - \alpha_1)}} \quad - \quad (36)$$

$$\phi_3^2 = \frac{m}{8} \; \frac{E_3^2 \pi^2}{n} \; \frac{1}{\alpha_1 \alpha_2 (\alpha_1 + \alpha_2)} \quad - \quad (37)$$

"*Numerical Example—Receiving Aerial.*—If at the base of the receiving aerial we find a current

$$\gamma_3 = 3 \,.\, 10^{-3} \text{ amp.} = 3 \,.\, 10^{-4} \text{ C.G.S.}$$

Then from (37) we have—

$$\phi_3 = 3 \,.\, 10^{-4} \,.\, 3 \,.\, 10^{6} \,.\, 16 \,.\, 10^{4} = 1.45 \,.\, 10^{8} \text{ C.G.S.} = 1.45 \text{ volt.}$$

"Applying results (32) to (36) we get—

$$E_3 = \frac{3 \,.\, 10^{-4} \,.\, 2\sqrt{2} \,.\, 3 \,.\, \sqrt{3} \,.\, 10^{9} \,.\, 16 \,.\, 10^{4}}{\sqrt{30} \,.\, 3.14}$$

$$= 28.4 \,.\, 10^{8} \text{ C.G.S.} = 28.4 \text{ volts.}$$

$$V'_3 = \frac{3.14}{2} \,.\, \frac{28.4 \,.\, 10^{8}}{0.3 - 0.06} = 185 \,.\, 10^{8} \text{ C.G.S.} = 185 \text{ volts.}$$

$$I'_3 = \frac{185 \,.\, 10^{8}}{48 \,.\, 10^{10}} = 3{,}8 \,.\, 10^{-2} \text{ C.G.S.} = 0{,}38 \text{ amp.}$$

$$V_{3\,max} = \frac{1}{2} 28.4 \,.\, 10^{8} \frac{3.14}{0.3} \left(\frac{0.06}{0.3}\right)^{\frac{0.06}{(0.3-0.06)}} = 104 \,.\, 10^{8} \text{ C.G.S.} = 104 \text{ volts.}$$

$$I_{3\,max} = \frac{104 \,.\, 10^{8}}{48 \,.\, 10^{10}} = 2.17 \,.\, 10^{-2} \text{ C.G.S.} = 0.217 \text{ amperes.}$$

"**The Secondary Circuit of the Receiving Station.**—The secondary of the receiving station consists essentially, as we have said, of the closed circuit of the condenser C_4. An E.M.F. e_4 is generated in it, at its point of coupling with the aerial, corresponding to the jigs transmitted by the sending station. Since this circuit is not interrupted by a spark-gap, it may be constructed so that its damping is exceedingly small; let us suppose it is $\alpha_4 = 0.006$.

"The jigs impressed on the circuit have the decrement $\alpha_1 = 0.06$ of the transmitting station, while in the receiving circuit itself the decrement is $\alpha_4 = 0.006$ as indicated above. At the commencement the resulting jig has obviously a zero value; it increases to a maximum and then decreases to zero again. Since α_1 is in our case much greater than α_4, the growth of the jig is much more rapid than its decline, hence we may assume, with an approximation which is good enough to give us at least the *maxima*, that we are dealing with an oscillation having a simple decrement α_4 and an amplitude equal to the maximum amplitude of the compound oscillation.

"The E.M.F. will be given by the formula—

$$e_4 = -M_{43}\frac{di_3}{dt} \qquad (38)$$

And the amplitude of it is—

$$E_4 = I_{3\,max}\,\pi n M_{43} \qquad (39)$$

As we said before, at the beginning of the action of the jig, the coherer behaves as a condenser. While it does so our equations will hold, but when it reaches a maximum, or when the coherer is cohered (which may be sooner), they have served their purpose and are not afterwards valid. We may therefore assume that (24), (25), (27), (28) remain valid, as they only deal with the instantaneous values of current and potential at a point in the circuit; they may therefore be transformed into the following:—

$$V'_4 = \frac{\pi E_4}{\alpha_1 - \alpha_2} \qquad (40)$$

$$V_{4\,max} = E_4\frac{\pi}{\alpha_4}\left(\frac{\alpha_1}{\alpha_4}\right)^{\frac{\alpha_1}{(\alpha_4-\alpha_1)}} \qquad (41)$$

$$I'_4 = \frac{E_4}{L_4 n(\alpha_1 - \alpha_4)} \qquad (42)$$

$$I_{4\,max} = E_4\frac{1}{nL_4\alpha_4}\left(\frac{\alpha_1}{\alpha_4}\right)^{\frac{\alpha_1}{(\alpha_4-\alpha_1)}} \qquad (43)$$

"*Numerical Example—Coherer Circuit.*—From (40) (41), and (6) we have, putting $M = 200$ and $L_4 - L_1 = 1656$,

$$E_4 = 2.17 \cdot 10^{-2} \cdot 3.14 \cdot 3 \cdot 10^6 \cdot 200 = 0.4 \text{ C.G.S.} = 0.4 \text{ volt.}$$

$$V'_4 = \frac{3.14 \cdot 0.4 \cdot 10^8}{0.06 - 0.006} = 23.6 \cdot 10^8 \text{ C.G.S.} = 23.6 \text{ volts.}$$

$$V_{4\,max} = 0.4 \cdot 10^8 \cdot \frac{3.14}{0.006}\left(\frac{0.06}{0.006}\right)^{-\frac{0.06}{0.054}} = 16.4 \cdot 10^8 \text{ C.G.S.}$$

$$= 16.4 \text{ volts.}$$

$$I'_4 = \frac{23.6 \cdot 10^8}{3 \cdot 10^6 \cdot 3.14 \cdot 1656} = 0.15 \text{ C.G.S.} = 1.5 \text{ ampere.}$$

$$I_{4\,max} = \frac{16.4 \cdot 10^8}{3 \cdot 10^6 \cdot 3.14 \cdot 1656} = 0.11 \text{ C.G.S.} = 1.1 \text{ ampere.}$$

"What we have given above may serve to give a quantitative idea, though only to the first order of approximation, of the phenomena of syntonic wireless telegraphy, and indicate the ways in which we may by calculation determine the dimensions of the various parts of an installation.

"The most notable result which arises from the numerical examples given is the demonstration of the difficulty of obtaining sufficiently high potentials for long-distance transmission.

"In general we may say the numbers resulting from the calculations are, if not always very close, at least sufficiently concordant to provide material for a study of the maxima and of the quantities dependent on them."

[*Note.*—The quantities on which the action of various types of detector depends are as follows:—

For coherers, V_{4max}.

„ magnetic detectors, I_{4max}.

„ thermo-galvanometers, γ^2.

J. E.-M.]

CHAPTER XXI.

TABLES AND NOTES.

IN order to render the volume of greater service to those who are directly interested in wireless telegraphy, some numerical and diagrammatic Tables, with other useful matter, are given in this chapter. Several of the Tables are new, having been specially calculated for the purpose; in other cases the source from which the Table is taken is noted.

MORSE CODES.

	The European or Continental Code.	American Morse Code.
A	- —	—
B	— - - -	— -
C	— - —	-
D	— -	—
E	-	
F	- —	—
G	— —	— —
H	- -	
I	- -	-
J	— — —	— - —
K	— - —	— —
L	- — -	——
M	— —	— —
N	—	—
O	— — —	
P	- — —	- -
Q	— — - —	—
R	—	
S	- -	-
T	—	—
U	—	—
V	- - —	- —
W	— —	— —
X	— - —	— -
Y	— — —	
Z	— — -	

NUMERALS.

	European	American
1	— — — —	- — — -
2	— — —	— -
3	- - — —	- —
4	- - - —	- - —
5	- -	— — —
6	— - - -	- - - -
7	— — -	— —
8	— — — -	— -
9	— — — —	— - —
0	— — — — —	— - - —
.	- - -	- — —
?	- - — — - -	— —
!	— — - — —	— — —

TABLE I.—DIELECTRIC STRENGTHS OF VARIOUS INSULATORS.

Kilovolts per centimetre required to break down the Insulator.

MATERIAL.	
Micanite -	4000
Mica -	2000
American Linen Paper, Paraffined -	540
Ebonite	538
Indiarubber -	492
Linseed Oil	83
Cotton Seed Oil -	67
Lubricating Oil -	48
Air Film, 2 mm. thick -	57
Air Film, 1.6 cm. thick - -	27

The numbers in this Table are taken from various sources.

TABLE II.—SPECIFIC INDUCTIVE CAPACITIES OR INDUCTIVITIES.

Inductivity of air taken as = 1.

Solids—	
Ebonite -	2.5
Glass, according to density	6.0 to 10.0
Guttapercha -	3.0
Indiarubber, pure -	2.2
" vulcanised	2.8
Mica - -	6.7
Solids—	
Paraffin wax -	2.1
Sulphur -	3.8 to 4.7
Liquids—	
Castor oil - -	4.8
Petroleum oil	2.1
Water at 15° C. -	80

NOTE.—For high frequency currents the inductivity is usually somewhat less than the figure given.

Definitions and Formulæ.—*Frequency* = Number of complete cycles per second: Symbol N. *Period* = Duration of one cycle: Symbol T. *Wave-length*, in meters $= \frac{3 \times 10^8}{N}$: Symbol λ. *Natural Frequency* of an open electric circuit $= \frac{5 \times 10^6}{\sqrt{CL}}$, where C = Capacity in microfarads, L = Inductance in centimetres.

Damping.—In a jig in which the oscillations are decreasing in amplitude the "damping" per half-period is the ratio of the amplitude of the second half-wave to the first. The Napierian logarithm of the reciprocal of this ratio is called the *logarithmic decrement:* Symbol δ. (See p. 210.)

TABLE III.—ELECTROSTATIC CAPACITY OF A SPHERE IN AIR AT A GREAT DISTANCE FROM OTHER CONDUCTORS.

Diameter.	Electrostatic Units.	In Millionths of a Microfarad (mmfd).
Cms.		
·5	.25	.278
1.0	·5	·556
1.5	·75	.831
2.0	1.0	1.11
2.5	1.25	1.39
3.0	1.5	1.66
4.0	2.0	2.22
5.0	2.5	2.78
10.0	5.0	5.56
100.0	50.0	55.60

(From formula, electrostatic capacity = radius.)

If the sphere is in another medium than air, multiply figure given by specific inductive capacity of medium.

TABLE IV.—CAPACITY OF PARALLEL PLATE CONDENSERS.

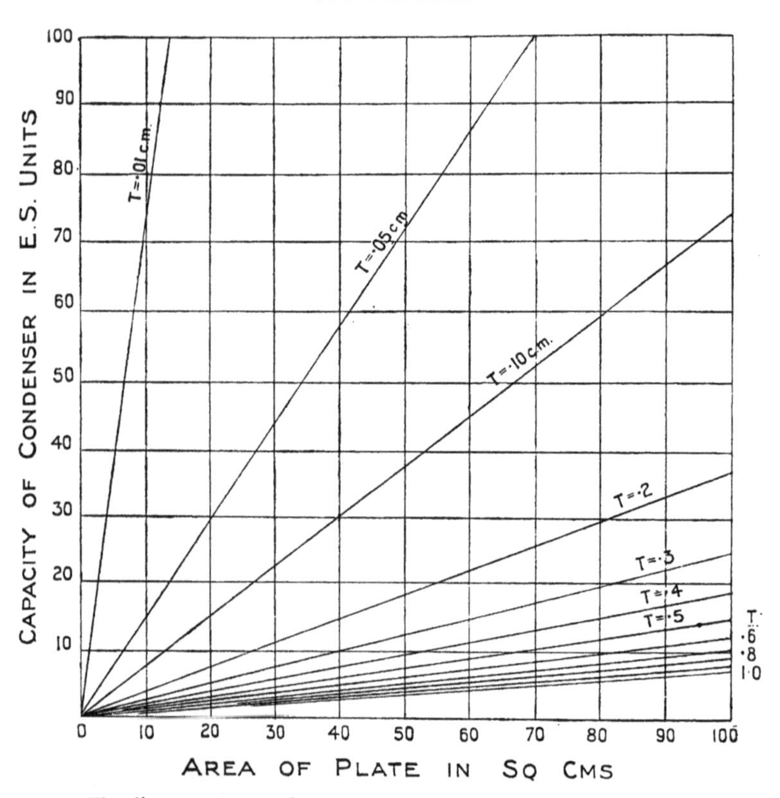

The lines correspond to plates at distances T=.01 cm., &c. &c. apart. The values are only roughly approximate, unless the area be great in proportion to the thickness T of the dielectric.

The dielectric is supposed to be air; if otherwise, multiply by inductivity.

To reduce to microfarads, divide E.S. units by 9×10^5.

TABLE V.—CAPACITY OF A CYLINDRICAL CONDENSER.

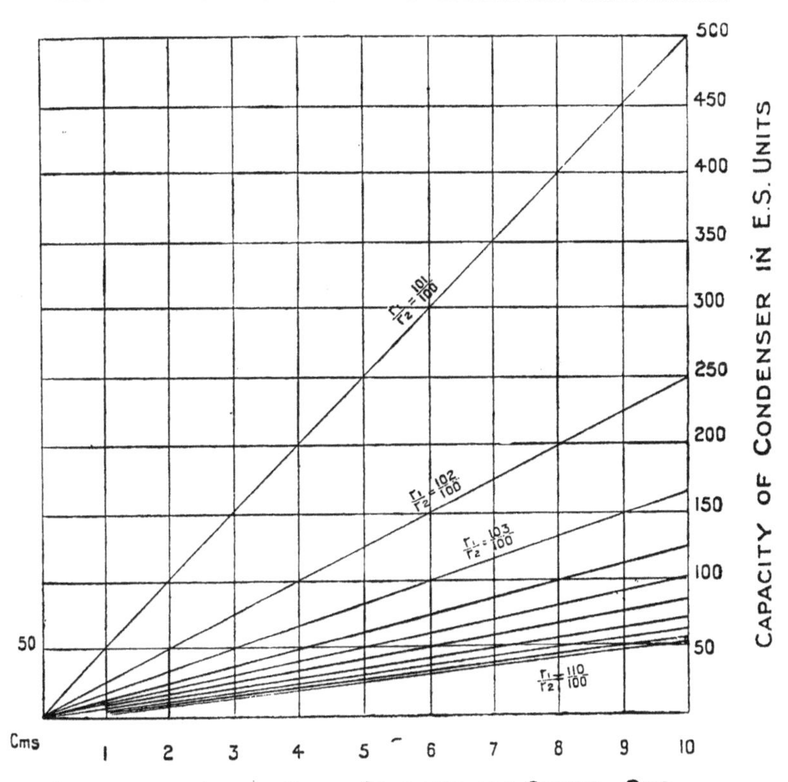

The dielectric is supposed to be air. Multiply by Spec. Ind. Capacity, if another dielectric is used.

The ratio r_1/r_2 is ratio of inside radius of outer conducting tube to outside radius of inner tube.

The lines between $r_1/r_2 = 103/100$ and $r_1/r_2 = 110/100$ are for values $r_1/r_2 = 104/100$, &c. &c. The absolute lengths of radii do not matter; only the ratio.

To obtain microfarads, divide E.S. units by 9×10^5.

TABLE VI.—CAPACITY OF VERTICAL WIRES OF VARIOUS DIAMETERS AND LENGTHS IN MILLIONTHS OF A MICROFARAD (mmfd.).

(From the formula, $K = l/(2 \log \frac{2l}{d})$)

No.	Diameter.	Height in Metres.					
S.W.G.	Cm.	5	10	20	30	40	50
4	·59	37·4	68.3	126	181	234	286
10	·325	34·4	63.7	118	170	220	269
16	.162	31.8	59.0	110	158	206	252
22	.071	29.1	54.2	102	147	191	234
36	.019	25.6	47.8	90.6	132	172	211

The capacity of a number of wires near one another is always much less than the sum of their individual capacities.

The capacity of a stranded wire is very little greater than that of a single wire of the same outside circumference.

TABLE VII.—INDUCTANCE OF HELIX PER CM. OF LENGTH.

D.	N=1.	N=2.	N=3.	N=4.
1	9.86	39·36	88.56	157
2	39·4	157	354·6	630
3	88.7	354·8	798.3	1419
4	157	630	1419	2520
5	246	984	2214	3936
6	354	1419	3186	5664
7	482	1928	4338	7712
8	630	2520	5670	10080
9	798	3192	7182	12768
10	986	3936	8856	15744

D = Diameter in cms.
N = Turns per cm.

The inductances are given in centimetres (1 henry = 10^9 cm.), and are calculated from the approximate formula $L = (\pi DN)^2$. To obtain approximate inductance of a helix of length l, multiply figure given in Table by l. Unless the helix is long in proportion to its diameter, the number given is only roughly approximate.

TABLE VIII.—RESISTANCE OF STRAIGHT COPPER WIRES TO CONSTANT AND TO HIGH FREQUENCY CURRENTS.

No. on S.W.G.	Diameter.		Resistance for Constant Current.	Resistance in Ohms per Meter for Uniform H.F. Currents of Frequencies		
				10^5.	10^6.	10^7.
	Inch.	Cm.	Ohms per Meter.			
12	.104	.264	.00312	.0109	.0328	.102
14	.080	.203	.00528	.0142	.0430	.130
16	.064	.163	.00824	.0180	.0527	.168
18	.048	.122	.0147	.0253	.0715	.223
20	.036	.091	.0260	.0358	.0933	.300
22	.028	.071	.0431	.0509	.1207	.389
30	.0124	.0315	.220	.220	.269	.900
32	.0108	.0274	.290	.290	.313	1.042
34	.0092	.0234	.398	.398	.416	1.235
36	.0076	.0193	.585	.585	.597	1.525
38	.0060	.0152	.943	.943	.950	2.080
40	.0048	.0122	1.466	1.466	1.473	2.580
42	.0040	.0102	2.109	2.109	2.200	3.260
44	.0032	.0081	3.300	3.300	3.300	4.070

Calculated from the formulæ of Kelvin, Maxwell, and Rayleigh as modified by Brylinski. If R be the C.C. resistance and R′ the H.F. resistance—

$$\frac{R'}{R}=1, \text{ when } \sqrt{a}=\sqrt{\frac{\mu\pi r^2 p}{\rho}}<0.5.$$

$$\frac{R'}{R}=0.079\times 4a-0.156\times 2\sqrt{a}+1.077\text{ ; when }\sqrt{a}\text{ between .5 and 1.5.}$$

$$\frac{R'}{R}=\frac{2\sqrt{2a}+1}{4}\text{ ; when }\sqrt{a}>1.5.$$

TABLE IX.—RATIO OF H.F. RESISTANCE R″ WITH DAMPING TO R′, THE RESISTANCE GIVEN IN TABLE VIII. FOR UNIFORM H.F. CURRENTS.

(From Dr E. H. Barton's formulæ.)

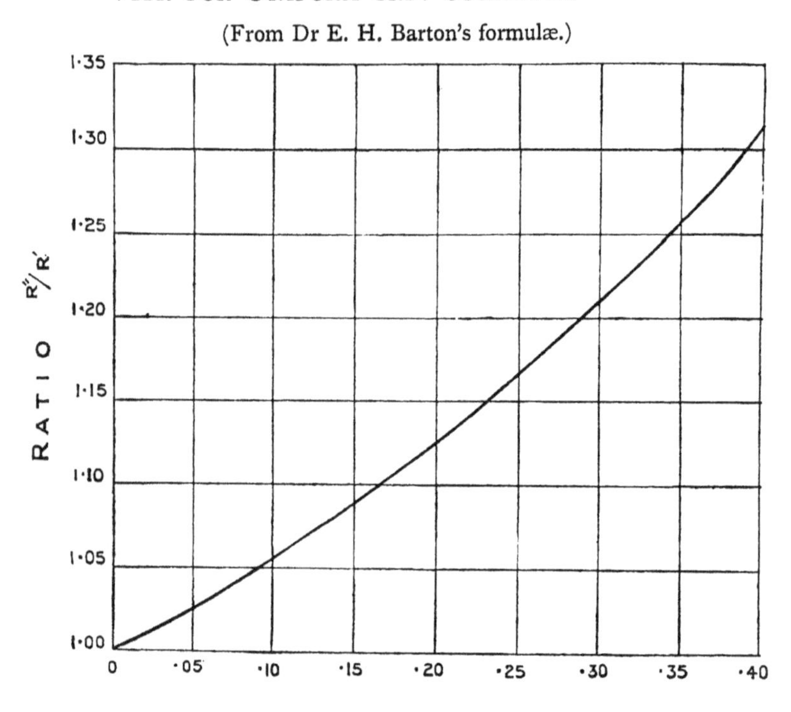

Damping factor = Logarithmic decrement per half-period divided by π.

Notes on Spark Resistance.

The resistance of the spark is in general the largest resistance in the path of the jig currents in wireless telegraph apparatus. It ranges from a fraction of an ohm for a spark length of about a millimetre up to 8 or 10 ohms for 1 cm., and its actual value depends on the materials of the electrodes, and also on the capacity in circuit and on the damping. The total resistance of a number of sparks in series is less than the resistance of a single spark whose length is the sum of the others; hence spark-gaps are now usually subdivided.

TABLE X.—Spark Lengths in Air, with Spark Balls of Various Sizes.

(From Mr E. Watson's Experiments, by kind permission of the Council of the Institute of Electrical Engineers.)

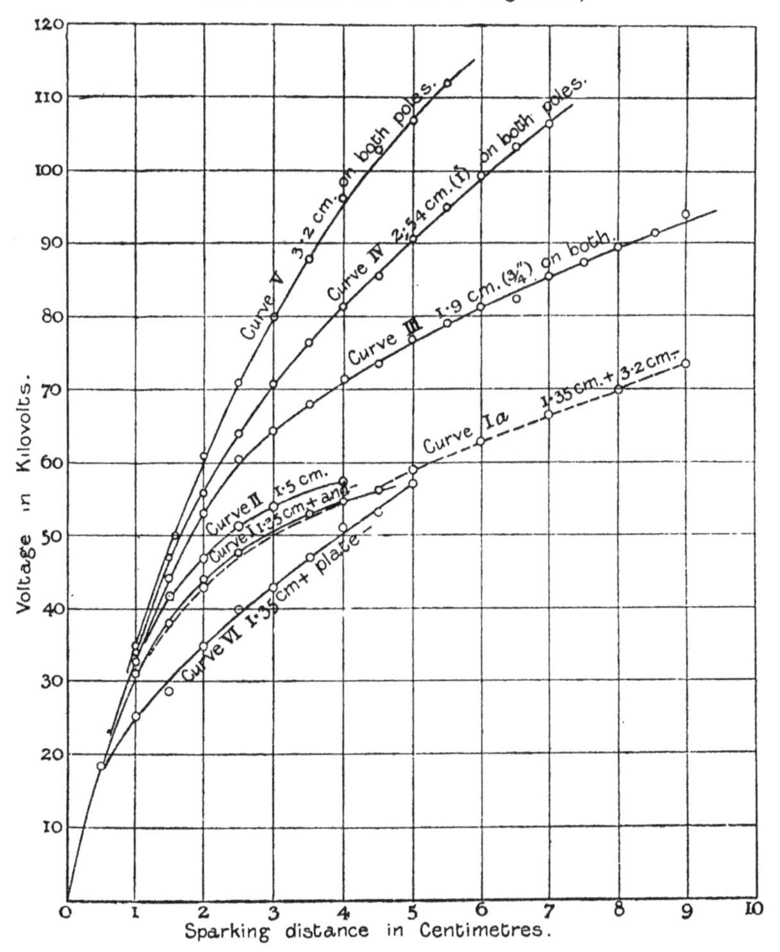

TABLE XI.—SMALL SPARKING DISTANCES IN AIR IN MILLIONTHS OF A MILLIMETRE.

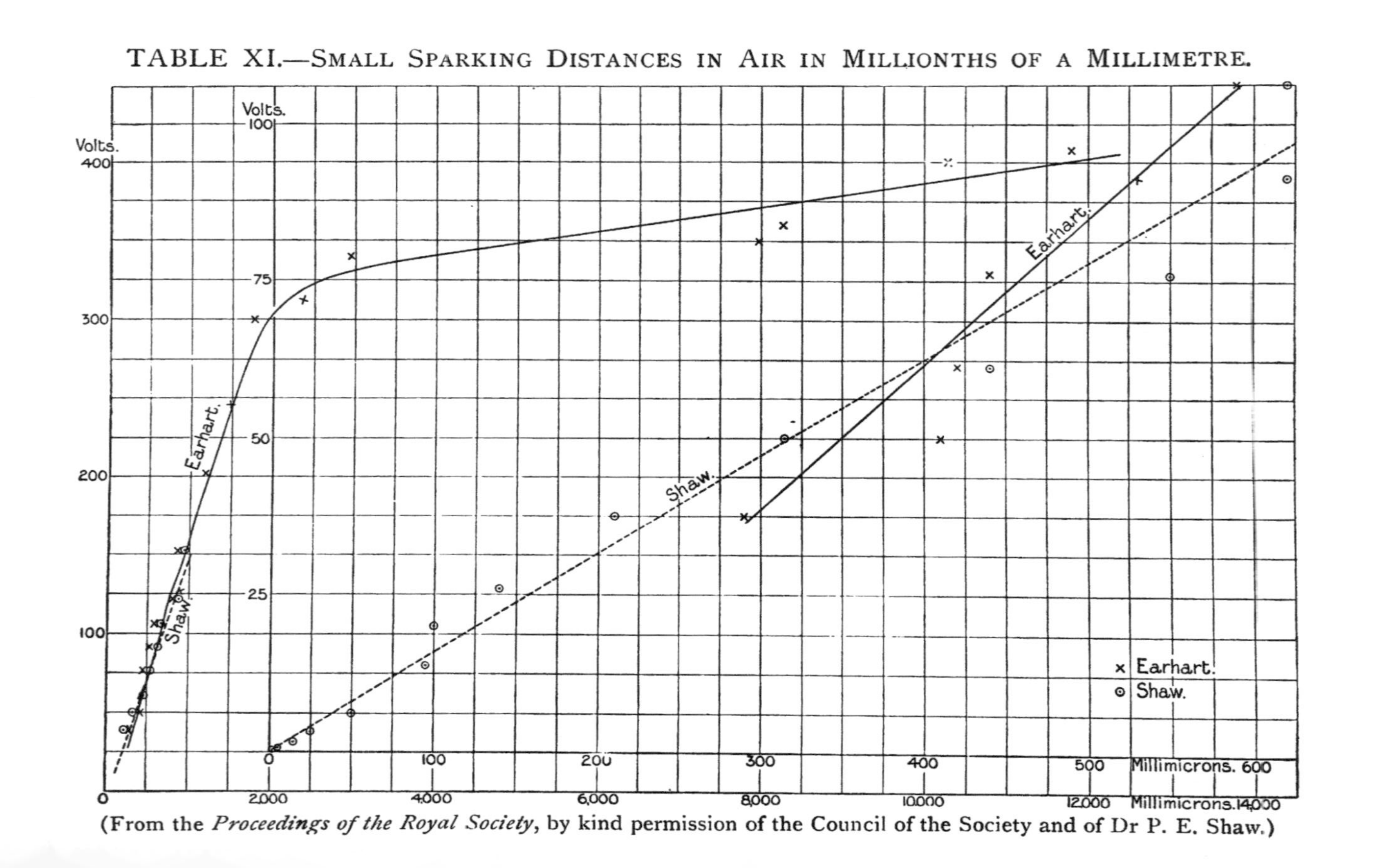

(From the *Proceedings of the Royal Society*, by kind permission of the Council of the Society and of Dr P. E. Shaw.)

Notes on Petrol Engines, Dynamos, and Accumulators.

Always pour your petrol into the tank through a strainer, for small solid particles may cause trouble in the carburettor.

Use a good quality of spirit, as (though almost anything will do in an emergency) you will very soon have trouble with the valves, &c., if you don't.

Study your engine till you know its weak points, and nurse them carefully.

If there is much sparking at the brushes of the dynamo, move them round the commutator until sparking is a minimum.

If the dark coloured marks usually called "flats" appear on the commutator, remove the brushes and polish with fine emery cloth, taking care that no emery gets into the bearings, finish with an oily rag, removing all dust.

Never discharge or charge your accumulators at greater rates than those stated by the makers. Never let the voltage of any cell drop below 1.85, and see that they are fully charged at least once a fortnight, even when not in use. Continue charging until the acid turns milky through the rapid evolution of gas from the plates. The density of the acid should be at least 1.195.

If the plates show signs of buckling, reduce the rates of charge and discharge, and examine them carefully from time to time in case short circuits should occur between adjacent plates.

Adhere strictly to any special instructions the makers may supply.

Notes on Ropes.

(From "THE GUNNER'S POCKET BOOK.")

Hemp Rope.—To calculate the working strain of rope, square the circumference in inches, and divide by 8 for the strain in tons.

To find the least size of rope to lift a given weight, multiply the weight in tons by 8 and extract the square root. The number found is the circumference in inches.

Wire Rope.—To find the safe strain for wire rope, multiply the square of the circumference in inches by .3 for iron, and .8 for steel wire. The breaking load is about three times the safe load.

Weight in lbs. per fathom is equal to the square of the circumference in inches. Thus 4-inch wire rope would weigh $4 \times 4 = 16$ lbs. per fathom

INDEX.

www.ingramcontent.com/pod-product-compliance
Ingram Content Group UK Ltd.
Pitfield, Milton Keynes, MK11 3LW, UK
UKHW040658180125
453697UK00010B/256